EMPLOYEE SAFETY MANUAL

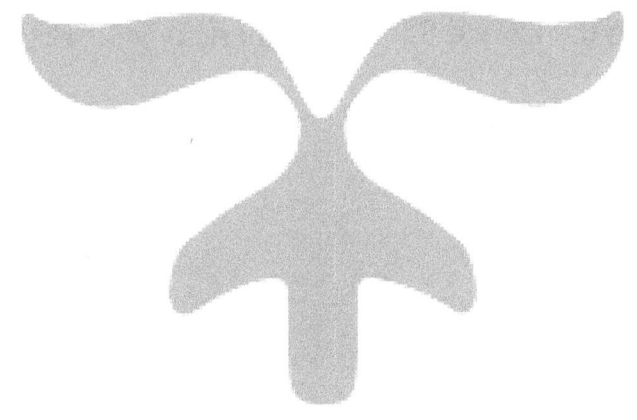

Printed in the United States of America

CONTENTS

1 CORPORATE SAFETY POLICY

It is the corporate policy that accident prevention shall be considered of primary importance in all phases of operation and administration. It is the intention of this organization's management to provide safe and healthy working conditions and always insist upon safe practices by all employees.

The personal safety and health of each employee is of primary importance. Prevention of occupational-induced injuries and illness is of such consequence that it will be given precedence over operating productivity. To the greatest degree possible, management will provide all mechanical and physical protection required for personal safety and health, but our employees must bear primary responsibility for working safely. A little common sense and caution can prevent most accidents from occurring.

This organization maintains a safety and health program conforming to the best practices of our field. To be successful, such a program must embody proper attitudes towards injury and illness prevention on the part of managers, supervisors, and employees.

It requires the cooperation in all safety and health matters, not only of the employer and employee, but between the employee and all co-workers. Only through such a cooperative effort can a safety program in the best interest of all be established and preserved.

2 GENERAL SAFETY RULES

Purpose

This organization recognizes the hazards in this industry and regards its employees to be of utmost importance. Our policies provide guidelines which shall govern job conditions and work practices for all employees. These will include the operation of equipment, handling of materials, and job conduct of employees.

It is our intent to furnish each employee a place of employment which is free from recognized hazards that are, or can be, responsible for injuries or disease. Each employee shall be required to comply with all identified safety and health standards and rules, regulations, and orders pursuant to our policy for prevention of accidents and industrial disease.

3   SAFETY PROGRAM GOALS

The objective of this organization is to provide a safety and health program that will reduce the number of injuries and illness to an absolute minimum, not merely in keeping with, but surpassing the best experience of similar operations by others. Our goal is zero accidents and injuries.

This guide is not intended to be an all-encompassing safety document. It is intended to set policy and to be used as a reference guide for common safety concerns encountered during normal operating/maintenance activities.

General Safety Rules

Failure to comply with the safety rules in this guide, as well as applicable federal, state, local, and customer rules, may be cause for termination. General safety rules include the following:

- No intoxicating beverages of any kind are permitted on the project.
- Horseplay, fighting, gambling, possession of firearms, drinking alcoholic beverages, using unauthorized drugs, or being under the influence of alcohol and/or drugs will not be tolerated.
- ALL INJURIES no matter how slight must be reported immediately to your supervisor.

- o An Accident/Injury/Illness Report must be completed.
- Smoke only in approved areas and dispose of butts in approved receptacles.
- Observe and follow all posted rules.
- All personally owned safety equipment such as hard hats or safety belts shall be approved by this organization before being used by an employee.
- Comply with all Safety Rules and Regulations at this organization.
- Eye protection equipment shall be worn when you are burning, chipping, welding, grinding, or drilling and when job site rules dictate.
- Orderliness must be maintained. Scrap, trash, and other wastes go in designated containers. Work areas must be cleaned up continually as the job progresses, with cords and hoses routed across walkways in a manner that will present no tripping hazard. All materials, tools, and equipment must be stored in a stable position (tied, stacked, or chocked) to prevent rolling or falling. A safe access way to all work areas must be maintained.
- Fall protection is mandatory when working at elevations four feet (6 feet in construction) or more above a lower level.
- Make sure you brace yourself properly when you prepare to pull on something that may be classed as heavy material.

- Make sure your grab is secure when pulling or tugging on heavy objects and that nothing will let go leaving you with no hold at all.
- Watch where you are walking. Running is not permitted anywhere, except in extreme emergency.
- Know the tools you are using - use the right tool for the right purpose.
- Employees are expected to report immediately any unsafe conditions including defective tools, equipment, or guards.
- Accidents can be prevented by using good judgment and being constantly alert.
- By using the right tools, practicing safe work methods, and not taking chances, you will minimize the potential for bodily injury to yourself and others.
- Be your brother's keeper. Consider what you do in terms of the hazard it may create for others.
- Ask your supervisor if you do not know or are in doubt as to the safe way of doing your job.
- Ignorance of safety rules and practices is no excuse for their violation.
- If you have a known handicap, such as diabetes, impaired sight, hearing, back trouble, hernia, heart trouble, aversion to heights and the like, ADVISE your supervisor so you will not be required to do a job that may injure you or someone around you.
- All unsafe conditions shall be reported to your supervisor

4   AERIAL LIFT SAFETY

Purpose

The purpose of this section is to outline policies and procedures for the safe operations of scissors lift and aerial lifts operated by this organization's employees. It applies to all locations that require employees to access elevated locations and/or use aerial work platforms.

Definitions

Aerial Lift - A piece of equipment, extendable and/or articulating, designed to position personnel and/or materials in elevated locations.

Lanyard - ANSI approved line designed for supporting one person, with one end connected to a safety harness and the other end attached to a suitable anchorage able to support 5,000 pounds of force. The anchorage can be a structural steel member, an approved lifeline, or other approved anchorage points.

Full Body Harness - ANSI approved body device designed for fall protection, which by reason of its attachment to a lanyard and safety line or an approved anchorage point, which will limit a fall to six feet or less.

Fall Protection

Full body harnesses and lanyards shall only be used, as intended by the manufacturer, for employee fall protection. Appropriate devices shall be used to provide 100% fall protection. The "D" ring on the body harness shall be positioned in the back up between the shoulder blades to minimize impact forces of the body in the event of a fall.

All fall protection equipment shall be carefully inspected prior to each use and periodically throughout the day. Safety equipment showing any signs of mildew, torn or frayed fabric or fiber, burns, excessive wear, or other damage or deterioration which could cause failure shall be permanently removed from service. All fall protection equipment shall be properly maintained and stored when not in use. This includes keeping dry and out of sunlight, away from caustics, corrosives or other materials that could cause defects.

Hard hats and safety harnesses shall be worn by employees in the bucket or platform of any aerial lift device. Other safety personal protective items may be required by either company or client safety policies. High visibility clothing is recommended while working in the air.

Equipment

Only devices approved for lifting personnel shall be used as aerial lifts. Loaders, forklifts, or other material lift devices shall NOT be used to transport employees to elevated locations nor as work platforms.

Modifications shall not be made to any aerial lift device without the expressed written authorization from the manufacturer. Buckets and bucket liners shall not be drilled, cut, welded on, etc.

Procedures

Lift equipment shall be inspected upon delivery to the jobsite, and daily prior to use. The daily inspection will include testing the controls prior to use, and all inspections shall be documented on the Aerial Lift Daily Inspection form.

Before extending or raising the boom or platform, outriggers (if so equipped), shall be positioned properly and the lift will be level. Outriggers shall be placed on mud mats or another SOLID surface and shall not be used to level the vehicle. If the lift is on unleveled ground, the wheels shall be chocked and the parking brake set. Sufficient clearance shall be checked before raising the lift.

Employees shall always keep both feet on the floor of the bucket or platform. When the lift must be moved, it shall only be moved when the bucket or platform is at the lowered position. For scissors lifts, this is lowered all the way down, and for aerial lifts, this is lowered to the lowest point that the operator can safely see to drive the vehicle.

Employees are required to wear full body safety harnesses with lanyards. The lanyards shall be attached to an engineered anchorage point inside the lift. Do NOT wrap the lanyard around a rail and tie back onto itself. Employees are NOT to anchor on structural members outside of the lift, unless exiting the lift to get on the structural members.

Platform lifts (scissors lifts) shall have a top and mid rail and a kick plate (toe board), along with an engineered anchorage point to tie off. Employees shall not climb nor stand on the mid or top rails, keeping both feet on the floor of the platform.

Tools, parts, or any materials shall not be dropped or thrown from the bucket or platform. When using welding or heating equipment from the bucket or platform, the vehicle shall be protected from sparks and slag and special care shall be taken to remove flammable objects away from the lifts.

Electrical Safety

Employees shall not position any aerial lifts closer than ten feet to a power line that carries up to 50 kilovolts. For each kilovolt over 50, add four inches.

Employees are to be trained concerning the hazards and precautions of working near power lines.

If the operator is unable to assess the clearances while operating the aerial lift, then a "spotter" must be used to observe the clearances and warn the operator.

Training

Aerial lift operators shall be trained and certified to use the various lifts on the jobsites.

5 BLOODBORNE PATHOGENS

Purpose

This organization is committed to providing a safe and healthful work environment for our entire staff. In pursuit of this endeavor, the following exposure control plan (ECP) is provided to eliminate or minimize occupational exposure to bloodborne pathogens.

This ECP includes:

- Determination of employee exposure.

- Implementation of various methods of exposure control, including:

  o Universal precautions.

  o Engineering and work practice controls.

  o Personal protective equipment; and

  o Housekeeping.

- Hepatitis B vaccination.

- Post-exposure evaluation and follow-up.

- Communication of hazards to employees and training.

- Recordkeeping; and

- Procedures for evaluating circumstances surrounding an exposure incident.

Administrative Duties

The site safety coordinator is responsible for the implementation of the ECP. The site safety coordinator will maintain, review, and update the ECP at least annually, and whenever necessary to include new or modified tasks and procedures.

Those employees who are determined to have occupational exposure to blood or other potentially infectious materials (OPIM) must comply with the procedures and work practices outlined in this ECP.

The site safety coordinator will maintain and provide all necessary personal protective equipment (PPE), engineering controls (e.g., sharps containers), labels, and red bags as required by the standard. Contacting the site safety coordinator will ensure that adequate supplies of the equipment are available in the appropriate sizes.

The site safety coordinator will be responsible for ensuring that all medical actions required are performed and that appropriate employee health and OSHA records are maintained.

The site safety coordinator will be responsible for training, documentation of training, and making the written ECP available to employees, OSHA, and NIOSH representatives.

Employee Exposure Determination

Part-time, temporary, contract, and per diem employees are covered by the standard. How the provisions of the standard will be met for these employees is described in this ECP, if applicable.

Methods of Implementation and Control

Universal Precautions

All employees will utilize universal precautions.

ECP

Employees covered by the bloodborne pathogens standard receive an explanation of this ECP during their initial training session. It will also be reviewed in their annual refresher training. All employees have an opportunity to review this plan at any time during their work shifts by contacting the site safety coordinator. If requested, we will provide an employee with a copy of the ECP free of charge and within 15 days of the request.

Engineering and Work Practice Controls

Engineering and work practice controls will be used to prevent or minimize exposure to bloodborne pathogens. This organization uses universal precautions.

This facility identifies the need for changes in engineering control and work practices through an annual review. We evaluate the need for new procedures or new products through an annual evaluation with potentially affected employees. The site safety coordinator will ensure effective implementation of these recommendations.

PPE

PPE is provided to each of our employees at no cost. Training is provided by the site safety coordinator in the use of the appropriate PPE for the tasks or procedures employees will perform. The types of PPE available to employees are as required by the universal precaution procedures.

Each employee using PPE must observe the universal precautions. Also follow the universal precautions for handling used PPE.

Housekeeping

Regulated waste is placed in containers that are closable, constructed to contain all contents and prevent leakage, appropriately labeled or color-coded, and closed prior to removal to prevent spillage or protrusion of contents during handling.

Bins and pails (e.g., wash or emesis basins) are cleaned and decontaminated as soon as feasible after visible contamination.

Hepatitis B Vaccination

The site safety coordinator will provide training to employees on hepatitis B vaccinations, addressing the safety, benefits, efficacy, methods of administration, and availability. The hepatitis B vaccination series is available at no cost after training and within 10 days of initial assignment to employees identified in the exposure determination section of this plan. Vaccination is encouraged unless documentation exists that the employee has previously received the series.

If an employee chooses to decline vaccination, the employee must sign a declination form. Employees who decline may request and obtain the vaccination later at no cost. Documentation of refusal of the vaccination is kept at the central office of each location.

Post-Exposure Evaluation and Follow-Up

Contact your immediate supervisor should an exposure incident occur.

An immediately available confidential medical evaluation and follow-up will be conducted by the designated heath care professional. Following the initial first aid (clean the wound, flush eyes, or other mucous membranes, etc.), accepted universal precaution response procedures will be utilized.

Administration of Post-Exposure Evaluation and Follow-up

The site safety coordinator ensures that the health care professional(s) responsible for employee's hepatitis B vaccination and post-exposure evaluation and follow-up are given a copy of OSHA's bloodborne pathogens standard. The site safety coordinator ensures that the health care professional evaluating an employee after an exposure incident receives all relative accident/incident data.

The site safety coordinator provides the employee with a copy of the evaluating health care professional's written opinion within 15 days after completion of the evaluation.

Procedures for Evaluating the Circumstances Surrounding an Exposure Incident

The site safety coordinator will review the circumstances of all exposure incidents to determine cause, corrective action implementation, and areas in need of procedural improvement. If it is determined that revisions need to be made, the site safety coordinator will ensure that appropriate changes are made to this ECP. The nature and date of the change(s) will be provided if any changes are made to the ECP.

BBP Employee Training

Each employee who has occupational exposure to bloodborne pathogens receives training conducted by the site safety coordinator. Our instructor(s) understands through education and/or experience the OSHA requirements of bloodborne pathogen hazards and controls.

Each employee who has occupational exposure to bloodborne pathogens receives training on the epidemiology, symptoms, and transmission of bloodborne pathogen diseases. In addition, the training program covers, at a minimum, all elements required to ensure potentially affected employees understand the hazard and controls of bloodborne pathogens. Training materials are available at the central office of each location.

6   CONFINED SPACE ENTRY

Roles and responsibilities

EH&S – Administers and approves confined space program and documents procedures and permits.
Entry Supervisor - Authorize and supervise permit-required confined space entry operations.
Entrants - Authorized to enter a permit space and do work.
Attendants - Stationed outside of a permit space to monitor entrant activity and perform duties listed on the permit.

Posting

All Permit-Required Confined Spaces must be labeled with the appropriate signage. EH&S department is responsible for posting signs on their Permit-Required Confined Spaces. If a space cannot be labeled, consult with EH&S about an alternative method of preventing entry into the space.

Permit Required Entry

A written Confined Space Entry Permit must be completed in full prior to all Permit-Required Confined Space entries. The Permit form can be used as a checklist for the job, along with forms for lockout/tagout and hot work procedures, if applicable. Upon conclusion of entry operations, the Entry Supervisor will cancel the Permit and file appropriately.

Procedures for Entry without Permit

Alternate Entry

Alternate entry procedures may be used for Permit-Required Confined Spaces where the only hazard is an actual or potentially hazardous atmosphere, which can be controlled with ventilation. Document using Hazardous Atmosphere Elimination to certify how the hazards were eliminated from the space, including date, location and signature of person making the determination. Post certification at space entrance. Follow procedures given in the Confined Space Entry Program.

Air monitoring parameters

Contaminant	Concentration
Carbon Monoxide (CO)	≤ 35 PPM
Hydrogen Sulfide (H2S)	≤ 10 PPM
Oxygen (O2)	$19.5 - 23.5\%$
Lower Explosive Limit (LEL)	$\leq 10\%$

The air should be tested at several levels in the space since gases may settle into layers. Continuous air monitoring should be done if the atmosphere can change, such as during welding, painting, descaling, cleaning with chemicals or working in sewers. Continuous air monitoring is recommended for all entries, taking readings every 15 minutes.

Training

Employees must be trained before performing any entries into any Permit-Required Confined Space. Employees will be trained in the duties of all positions involved in the entry.

The company must provide training on specific procedures and safety precautions related to the confined spaces in their areas, and when certain work is done in a confined space.

Employees must receive periodic refresher training and additional training anytime there is a change in assignment, operation, or procedures.

All confined space training will be documented with the date of training and a listing of trainees.

Emergency Procedures and Rescue Services

Rescue Services

Rescue involving confined space entry will **not** be performed by the host company employees.

The Fire Department will provide confined space emergency rescue services at the company's facilities. The following requirements must be met:

- Responsible person for the entry must determine if the resources of the Fire Department are adequate for the entry being performed.
- The Fire Department must evaluate the entry conditions and space to determine if their rescue service is within the scope of what is needed for safe rescue from the space.

Self-Rescue

Entrants must Self-Rescue if they feel ill, are injured, detect a problem or prohibited condition or if directed to leave the space. An entrant should have the foresight and ability to remove themselves from the hazard before it becomes a bigger problem requiring Assisted or Entry Rescue.

Non-Entry Rescue (or Assisted Rescue)

When an entrant is not capable of self-rescue, the attendant assists removing the entrant from the space using a mechanical retrieval device that is attached to the entrant. This must only be done from outside the space and if it is safe. A vertical confined space more than 5 feet deep requires a mechanical retrieval device and entrants to wear a chest or full-body harness attached to a retrieval line.

Supervisor and Attendant Responsibilities during Entry Rescue

- Provide the rescue service with information on the work being done and any chemicals in use or other hazardous atmosphere producing activities.
- Provide the Entry Permit to rescue service personnel.
- Provide rescue service with any observations or information about the emergency.
- Keep unauthorized personnel out of the area.

Safe Air Limits for Confined Space Entry:

Oxygen: Between 19.5% and 23.5%
LEL: Less than 10% of known LEL (LEL = Lower Explosive Limit)
H2S: Less than 10 ppm
CO: Less than 35 ppm

If the air quality inside the confined space is safe with no fresh air ventilation, and if the only hazard is an actual or potentially hazardous atmosphere that can be controlled by continuous forced air ventilation, the space may be reclassified temporarily as an "Alternate Entry Procedures" space.

Re-evaluate space and/or modify entry procedure before re-entry.

7 CRANE & HOIST SAFETY

Purpose

Many types of cranes, hoists, and rigging devices are used for lifting and moving materials. This organization's policy is to maintain a safe workplace for its employees; therefore, it cannot be overemphasized that only qualified individuals shall operate these devices.

The safety rules and guidance in this chapter apply to all operations that involve the use of cranes and hoists installed in or attached to buildings and to all employees, supplemental labor, and subcontractor personnel who use such devices.

Responsibilities

Supervisors are responsible for:

- Ensuring that employees under their supervision receive the required training and are qualified to operate the cranes and hoists in their areas.
- Providing training for prospective crane and hoist operators. This training must be conducted by a qualified instructor.
- Ensuring that hoisting equipment is inspected and tested monthly by a responsible individual and that rigging equipment is inspected annually.

Crane and Hoist Operators are responsible for:

- Operating hoisting equipment safely.
- Conducting functional tests prior to using the equipment.
- Selecting and using rigging equipment appropriately.

Maintenance Department is responsible for:

- Performing annual maintenance and inspection of all cranes and hoists.
- Conducting periodic and special load tests of cranes and hoists.
- Maintaining written records of inspections and tests and providing copies of all inspections and test results to facility managers.
- Inspecting and load testing cranes and hoists following modification or extensive repairs (e.g., a replaced cable or hook, or structural modification.)
- Scheduling a non-destructive test and inspection for crane and hoist hooks at the time of the periodic load test, and testing and inspecting before use new replacement hooks and other hooks suspected of having been overloaded.
- Maintaining all manuals for cranes and hoists in a central file for reference.

Safety Coordinator is responsible for:

- Arranging training for all Crane & Hoist Operators
- Periodically verifying monthly test and inspection reports.
- Interpreting crane and hoist safety rules and standards.

Crane and Hoist Safety Rules

Operators shall comply with the following rules while operating the cranes and hoists:

- Do not engage in any practice that will divert your attention while operating the crane.
- Respond to signals only from the person who is directing the lift or any appointed signal person. Always obey a stop signal, no matter who gives it.
- Do not move a load over people. People shall not be placed in jeopardy by being under a suspended load. Also, do not work under a suspended load unless the load is supported by blocks, jacks, or a solid footing that will safely support the entire weight. Have a crane or hoist operator remain at the controls or lock open and tag the main electrical disconnect switch.
- Ensure that the rated load capacity of a crane's bridge, individual hoist, or any sling or fitting is not exceeded. Know the weight of the object being lifted or use a dynamometer or load cell to determine the weight.
- Check that all controls are in the OFF position before closing the main line disconnect switch.

- If spring-loaded reels are provided to lift pendants clear off the work area, ease the pendant up into the stop to prevent damaging the wire.
- Avoid side pulls. These can cause the hoist rope to slip out of the drum groove, damaging the rope or destabilizing the crane or hoist.
- To prevent shock loading, avoid sudden stops or starts. Shock loading can occur when a suspended load is accelerated or decelerated and can overload the crane or hoist. When completing an upward or downward motion, ease the load slowly to a stop.

Crane and Hoist Operating Rules

Pre-operational Test

At the start of each work shift, operators shall do the following steps before making lifts with any crane or hoist:

- Test the upper-limit switch. Slowly raise the unloaded hook block until the limit switch trips.
- Visually inspect the hook, load lines, trolley, and bridge as much as possible from the operator's station; in most instances, this will be the floor of the building.
- If provided, test the lower-limit switch.
- Test all direction and speed controls for both bridge and trolley travel.

- Test all bridge and trolley limit switches, where provided, if operation will bring the equipment in close proximity to the limit switches.
- Test the pendant emergency stop.
- Test the hoist brake to verify there is no drift without a load.
- If provided, test the bridge movement alarm.
- Lock out and tag for repair of any crane or hoist that fails any of the above tests.

Moving a Load

- Center the hook over the load to keep the cables from slipping out of the drum grooves and overlapping, and to prevent the load from swinging when it is lifted. Inspect the drum to verify that the cable is in the grooves.
- Use a tag line when loads must traverse long distances or must otherwise be controlled. Manila rope may be used for tag lines.
- Plan and check the travel path to avoid personnel and obstructions.
- Lift the load only high enough to clear the tallest obstruction in the travel path.
- Start and stop slowly.
- Land the load when the move is finished. Choose a safe landing.
- Never leave suspended loads unattended. In an emergency where the crane or hoist has become

inoperative, if a load must be left suspended, barricade and post signs in the surrounding area, under the load, and on all four sides. Lock open and tag the crane or hoist's main electrical disconnect switch.

Parking a Crane or Hoist

- Remove all slings and accessories from the hook. Return the rigging device to the designated storage racks.
- Raise the hook at least 2.1 m (7-ft) above the floor.
- Store the pendant away from aisles and work areas or raise it at least 2.1 m (7 ft) above the floor.
- Place the emergency stop switch (or push button) in the OFF position.

Rigging

General Rigging Safety Requirements

- Only select rigging equipment that is in good condition. All rigging equipment shall be inspected annually; defective equipment is to be removed from service and destroyed to prevent inadvertent reuse. The load capacity limits shall be stamped or affixed to all rigging components.
- Company policy requires a minimum safety factor of 5 to be maintained for wire rope slings.

The following types of slings shall be rejected or destroyed:

- Nylon slings with
- Abnormal wear.
- Torn stitching.
- Broken or cut fibers.
- Discoloration or deterioration.
- Wire-rope slings with
- Kinking, crushing, bird caging, or other distortions.
- Evidence of heat damage.
- Cracks, deformation, or worn end attachments.
- Six randomly broken wires in a single rope lay.
- Three broken wires in one strand of rope.
- Hooks opened more than 15% at the throat.
- Hooks twisted sideways more than 10deg. from the plane of the unbent hook.
- Alloy steel chain slings with
- Cracked, bent, or elongated links or components.
- Cracked hooks.
- Shackles, eye bolts, turnbuckles, or other components that are damaged or deformed.

Rigging a Load

Do the following when rigging a load:

- Determine the weight of the load. Do not guess.
- Determine the proper size for slings and components.
- Do not use manila rope for rigging.

- Make sure that shackle pins and shouldered eyebolts are installed in accordance with the manufacturer's recommendations.
- Make sure that ordinary (shoulder less) eyebolts are threaded in at least 1.5 times the bolt diameter.
- Use safety hoist rings (swivel eyes) as a preferred substitute for eye bolts wherever possible.
- Pad sharp edges to protect slings. Remember that machinery foundations or angle-iron edges may not feel sharp to the touch but could cut into rigging when under several tons of load. Wood, tire rubber, or other pliable materials may be suitable for padding.
- Do not use slings, eyebolts, shackles, or hooks that have been cut, welded, or brazed.
- Install wire-rope clips with the base only on the live end and the U-bolt only on the dead end. Follow the manufacturer's recommendations for the spacing for each specific wire size.
- Determine the center of gravity and balance the load before moving it.
- Initially lift the load only a few inches to test the rigging and balance.

Crane Overloading

Cranes or hoists shall not be loaded beyond their rated capacity for normal operations. Any crane or hoist suspected of having been overloaded shall be removed from service by locking open and tagging the main disconnect switch. Additionally, overloaded cranes shall be inspected, repaired, load tested, and approved for use before being returned to service.

Working at Heights on Cranes or Hoists

Anyone conducting maintenance or repair on cranes or hoists at heights greater than 1.8 m (6 ft) shall use fall protection. Fall protection should also be considered for heights less than 1.8 m. Fall protection includes safety harnesses that are fitted with a lifeline and securely attached to a structural member of the crane or building or properly secured safety nets.

Use of a crane as a work platform should only be considered when conventional means of reaching an elevated worksite are hazardous or not possible.

Inspection, Maintenance, and Testing

All tests and inspections shall be conducted in accordance with the manufacturer's recommendations.

Monthly Tests and Inspections

- All in-service cranes and hoists shall be inspected monthly, and the results documented.
- Defective cranes and hoists shall be locked and tagged "out of service" until all defects are corrected. The inspector shall initiate corrective action by notifying the facility manager or building coordinator.

Annual Inspections

The annual PM and inspection shall cover

- Hoisting and lowering mechanisms.
- Trolley travel or monorail travel.
- Bridge travel.
- Limit switches and locking and safety devices.
- Structural members.
- Bolts or rivets.
- Sheaves and drums.
- Parts such as pins, bearings, shafts, gears, rollers, locking devices, and clamping devices.
- Brake system parts, linings, pawls, and ratchets.
- Load, wind, and other indicators over their full range.
- Gasoline, diesel, electric, or other power plants.
- Chain-drive sprockets.
- Crane and hoist hooks.
- Electrical apparatus such as controller contractors, limit switches, and push button stations.
- Wire rope.
- Hoist chains.

8 ELECTRICAL SAFETY

Purpose

The Electrical Safety program is designed to prevent electrically related injuries and property damage. This program also provides for proper training of maintenance employees to ensure they have the requisite knowledge and understanding of electrical work practices and procedures.

Only employees qualified in this program may conduct adjustment, repair or replacement of electrical components or equipment.

Responsibilities

Management

- Provide training for qualified and unqualified employees.
- Conduct inspections to identify electrical safety deficiencies.
- Guard and correct all electrical deficiencies promptly.
- Ensure all new electrical installations meet codes and regulations.

Employees

- Report electrical deficiencies immediately.
- Not work on electrical equipment unless authorized and trained.

- Properly inspect all electrical equipment prior to use.

Hazard Control

Engineering Controls

- All electrical distribution panels, breakers, disconnects, switches, junction boxes shall be completely enclosed.
- Watertight enclosure shall be used where there is possibility of moisture entry either from operations or weather exposure.
- Electrical distribution areas will be guarded against accidental damage by locating in specifically designed rooms, use of substantial guard posts and rails and other structural means.
- A clear approach and 3-foot side clearance shall be maintained for all distribution panels.
- All conduits shall be fully supported throughout its length. Non-electrical attachments to conduit are prohibited.
- All non-rigid cords shall be provided strain relief where necessary.

Administrative Controls

- Only trained and authorized employees may conduct repairs to electrical equipment.
- Contractors performing electrical work must be hold a license for the rated work.

- Areas under new installation or repair will be sufficiently guarded with physical barriers and warning signs to prevent unauthorized entry.
- Access to electrical distribution rooms is limited to those employees who have a need to enter.
- All electrical control devices shall be properly labeled.
- Work on energized circuits is prohibited unless specifically authorized by NextG Networks management.
- All qualified employees will follow established electrical safety procedures and precautions.

Protective Equipment

- Qualified employees will wear electrically rated safety shoed/boots.
- All tools used for electrical work shall be properly insulated.
- Electrical rated gloves shall be available for work on electrical equipment.
- Electrically rated matting will be installed in front of all distribution panels in electric utility rooms.

Electrical Equipment

Examination

- Electrical equipment shall be free from recognized hazards that are likely to cause death or serious physical harm to employees. Safety of equipment shall be determined using the following considerations:
- Suitability for installation and use in conformity with the provisions of this subpart. Suitability of equipment for an identified purpose may be evidenced by listing or labeling for that identified purpose.
- Mechanical strength and durability, including, for parts designed to enclose and protect other equipment, the adequacy of the protection thus provided.
- Electrical insulation.
- Heating effects under conditions of use.
- Arcing effects.
- Classification by type, size, voltage, current capacity, and specific use.
- Other factors which contribute to the practical safeguarding of employees using or likely to come in contact with the equipment.

Identification of Disconnecting Means and Circuits

- Each disconnecting means for motors and appliances shall be legibly marked to indicate its purpose. Each service, feeder, and branch circuit, at its disconnecting means or over current device, shall be legibly marked to indicate its purpose. These markings shall be of sufficient durability to withstand the environment involved.

- A disconnecting means is a switch that is used to disconnect the conductors of a circuit from the source of electric current. Disconnect switches are important because they enable a circuit to be opened, stopping the flow of electricity, and thus can effectively protect workers and equipment.

- Each disconnect switch or over current device required for a service, feeder, or branch circuit must be clearly labeled to indicate the circuit's function, and the label or marking should be located at the point where the circuit originates. For example, on a panel that controls several motors or on a motor control center, each disconnect must be clearly marked to indicate the motor to which each circuit is connected.

- All labels and markings must be durable enough to withstand weather, chemicals, heat, corrosion, or any other environment to which they may be exposed.

Definition of Terms

- Qualified Worker: An employee trained and authorized to conduct electrical work.
- Unqualified: Employees who have not been trained or authorized by management to conduct electrical work.

Training

Training for Unqualified Employees

Training for Unqualified Employees is general electrical safety precautions to provide an awareness and understanding of electrical hazards.

Electrical Safety Rules for Non-Qualified Workers

- Do not conduct any repairs to electrical equipment.
- Report all electrical deficiencies to your supervisor.
- Do not operate equipment if you suspect and electrical problem.
- Water and electricity do not mix.
- Even low voltages can kill or injure you
- Do not use cords or plugs if the ground prong is missing.
- Do not overload electrical receptacles.

Training for Qualified Employees

Training for Qualified Employees includes specific equipment procedures and requirements of Electrical Safety, 29 CFR 1910.331 to 1910.339

Training for employees (qualified and unqualified) who face a risk of electric shock that is not reduced to a safe level by proper electrical installation.

Training can be either in the classroom or on-the-job type.

The degree of training required is dependent upon the risk to the employee.

Specific required training includes:

- Skills and techniques necessary to distinguish exposed live parts from other parts of electric equipment.
- Skills and techniques necessary to determine the nominal voltage of exposed live parts,
- Clearance distances specified in OSHA Standard 1910.333(c) and the corresponding voltages to which the qualified person will be exposed.

9 EMERGENCY ACTION PLAN

Purpose

The purpose of this organization's Emergency Action Plan is to comply with the Occupational Safety and Health Administration's (OSHA's) Emergency Action Plan Standard, 29 CFR 1910.38, and to prepare employees for dealing with emergency situations.

This plan is designed to minimize injury and loss of human life and company resources by training employees, procuring, and maintaining necessary equipment, and assigning responsibilities. This plan applies to all emergencies that may reasonably be expected to occur at all locations.

Assignment of Responsibility

Emergency Plan Manager

The designated employee shall manage the Emergency Action Plan for this organization The Emergency Plan Manager shall also maintain all training records pertaining to this plan. The plan manager is responsible for scheduling routine tests of this organization's emergency notification system with the appropriate authorities.

The Emergency Plan Manager shall also coordinate with local public resources, such as fire department and emergency medical personnel, to ensure that they are prepared to respond as detailed in this plan.

Emergency Plan Coordinator

The Emergency Plan Coordinators are responsible for instituting the procedures in this plan in their designated areas in the event of an emergency. Note: Coordinators may also be given the responsibility of accounting for employees/visitors after an evacuation has occurred.

Management

This organization will provide adequate controls and equipment that, when used properly, will minimize or eliminate risk of injury to employees in the event of an emergency. This organization's management will ensure proper adherence to this plan through regular review.

Supervisors

Supervisors shall follow and ensure that their employees are trained in the procedures outlined in this plan.

Employees

Employees are responsible for following the procedures described in this plan.

Contractors/Vendors

Contract employees are responsible for complying with this plan.

Plan Implementation

Reporting Fire and Emergency Situations

All fires and emergency situations will be reported as soon as possible to the Emergency Plan Coordinator by one of the following means:

- Verbally as soon as possible during normal work hours; or

- By telephone if after normal work hours or on weekends.

To eliminate confusion and the possibility of false alarms, only the Emergency Plan Coordinator is authorized to contact the appropriate community emergency response personnel.

Under no circumstances shall an employee attempt to fight a fire that has passed the incipient stage (that which can be put out with a fire extinguisher), nor shall any employee attempt to enter a burning building to conduct search and rescue.

These actions shall be left to emergency services professionals who have the necessary training, equipment, and experience (such as the fire department or emergency medical professionals). Untrained individuals may endanger themselves and/or those they are trying to rescue.

Informing this organization's employees of fires and emergency situations

In the event of a fire or emergency, the Emergency Plan Coordinator shall ensure that all employees are notified as soon as possible using the building alarm system (which includes both audible and visual alarms 24 hours a day). The Emergency Plan Coordinator shall provide special instructions to all employees.

If a fire or emergency occurs after normal business hours, the Emergency Plan Coordinator shall contact all employees not on shift of future work status, depending on the nature of the situation.

Emergency Contact Information

The Emergency Plan Coordinator shall maintain a list of all employees' personal emergency contact information and shall keep the list in a designated location for easy access in the event of an emergency.

Evacuation Routes

Emergency evacuation escape route plans are posted throughout each location. If a fire/emergency alarm is sounded or instructions for evacuation are given by the Emergency Plan Coordinator, all employees shall immediately exit the building(s) at the nearest exits as shown in the escape route plans and shall meet as soon as possible at the designated assembly location. Employees with offices shall close the doors (unlocked) as they exit the area.

Mobility impaired employees and their assigned assistants will gather at the designated assembly location within the building to ensure safe evacuation.

Securing Property and Equipment

If evacuation of the premises is necessary, some items may need to be secured to prevent further detriment to the facility and personnel on hand (such as securing confidential/irreplaceable records and/or shutting down equipment to prevent release of hazardous materials).

All individuals remaining behind to shut down critical systems or utilities shall be capable of recognizing when to abandon the operation or task. Once the property and/or equipment have been secured, or the situation becomes too dangerous to remain, these individuals shall exit the building by the nearest escape route as soon as possible and meet the remainder of the employees at the designated assembly location.

Advanced Medical Care

Under no circumstances shall an employee provide advanced medical care and treatment. These situations shall be left to emergency services professionals, or Emergency Plan Coordinator, who have the necessary training, equipment, and experience. Untrained individuals may endanger themselves and/or those they are trying to assist.

Accounting for Employees/Visitors After Evacuation

Once an evacuation has occurred, the Emergency Plan Coordinator shall account for each employee/visitor assigned to them at the designated assembly location. Each employee is responsible for reporting to the appropriate Emergency Plan Coordinator so an accurate head count can be made. All employee counts shall then be reported to the Emergency Action Plan Manager as soon as possible.

Re-Entry

Once the building has been evacuated, no one shall re-enter the building for any reason, except for designated and trained rescue personnel (such as fire department or emergency medical professionals). Untrained individuals may endanger themselves and/or those they are trying to rescue.

All employees shall remain at the designated assembly location until the fire department or other emergency response agency notifies Emergency Plan Coordinator that either:

- The building is safe for re-entry, in which case personnel shall return to their workstations; or

- The building/assembly area is not safe, in which case personnel shall be instructed by Emergency Plan Coordinator on how/when to vacate the premises.

Sheltering in Place

If chemical, biological, or radiological contaminants are released into the environment in such quantity and/or proximity, authorities and/or the Emergency Plan Coordinator may determine that is safer to remain indoors rather than to evacuate employees. The Emergency Action Plan Manager shall announce sheltering in place status.

Severe Weather

The Emergency Action Plan Manager shall announce severe weather alerts (such as tornados) by public address system or other means of immediate notification available at each Branch location. All employees shall immediately retreat to the **designated assembly location** until the threat of severe weather has passed as communicated by the Emergency Action Plan Manager.

Emergency Action Plan Training

Employee Training

All employees shall receive instruction on the Emergency Action Plan as part of new employee orientation upon hire. Additional training shall be provided:

- When there are any changes to the plan and/or facility.

- When an employee's responsibilities change; and

- Annually as refresher training.

Items to be reviewed during the training include:

- Proper housekeeping.
- Fire prevention practices.
- Fire extinguisher locations, usage, and limitations.
- Threats, hazards, and protective actions.
- Means of reporting fires and other emergencies.
- Names of Emergency Action Plan Manager and coordinators.

- Individual responsibilities.
- Alarm systems.
- Escape routes and procedures.
- Emergency shut-down procedures.
- Procedures for accounting for employees and visitors.
- Closing doors.
- Sheltering in place.
- Severe weather procedures; and
- Emergency Action Plan availability.

Fire/Evacuation Drills

Fire/evacuation drills shall be conducted at least annually in coordination with local police and fire departments. Additional drills shall be conducted if physical properties of the business change, processes change, or as otherwise deemed necessary.

Plan Evaluation

The Emergency Action Plan shall be reviewed annually, or as needed if changes to the worksite are made, by the Director of Human Resources. Following each fire drill, the Director of Human Resources and Emergency Plan Coordinator shall evaluate the drill for effectiveness and weaknesses in the plan and shall implement changes to improve it.

10 FALL PROTECTION

Purpose

To comply with or exceed OSHA standards, and to set company policy regarding fall protection.

Affected Employees

Each employee on a walking/working surface with an unprotected side or edge that is four feet or more above a lower level shall be protected from falling.

Types of Protection

Primary Fall Protection Systems

- Primary fall protection systems provide walking/working surfaces in elevated areas to be free from unprotected sides or edges (including floor openings). Primary fall protection systems can be accomplished by placing guardrails on unprotected sides or edges and by covering any floor openings.
- Guardrail systems shall have a top rail height of approximately 42 inches. The mid-rail height shall be approximately 21 inches and the systems shall be equipped with a 4-inch toe board whenever necessary. Guardrail systems shall be capable of withstanding, without failure, a force of at least 200 pounds, applied within 2 inches of the top edge, in any outward or downward direction, at any point along the top edge.

- Floor opening covers shall be capable of supporting the maximum potential load they may be subject to. They must completely cover the opening and be secured to prevent accidental displacement.

NOTE: Primary fall protection systems shall be used whenever possible and shall always be given first consideration. However, secondary fall protection may be necessary during the installation of primary systems.

Secondary Fall Protection Systems

- A secondary fall protection system consists of a full body harness and two positive locking shock absorbing lanyards. This system shall be used when employees are exposed to a fall of 4 feet or more and primary systems are not possible.
- Only full body safety harnesses and positive locking shock absorbing lanyards capable of supporting 5000 pounds shall be used. Safety belts shall not be used for fall protection.
- The tie off point must be at waist level or higher and have the capabilities to support no less than 5000 pounds.

Lifelines

Lifelines are points of attachment for fall protection. The following will apply:

- Lifeline systems must be capable of supporting 5,000 pounds for each person attached to it.
- Lifelines may be either horizontal or vertical.
- Lifelines shall be protected against being cut or abraded.
- Horizontal lifelines must be made of at least 3/8-inch wire rope properly supported to withstand at least 5000 pounds impact and pulled tight enough to prevent deflection.
- Vertical lifelines shall be made of either 5/8- or 3/4-inch nylon rope and equipped with an approved rope grab, lanyard, and safety harness.
- Self-retracting lifelines must limit free fall distances to two feet or less and must be capable of supporting 5,000 pounds.
- All equipment must be inspected, by the user, prior to each use.
- Defective equipment shall be removed from service at once, to prevent any use.

NOTE: Secondary systems may still be necessary even when primary systems are in place. Examples of when safety harnesses are needed, even when guardrails are in place, include areal lifts, suspension scaffolds, and man cages.

Other Systems

Other fall protection systems such as safety monitoring, safety net, and warning line systems must be approved by the Safety Coordinator before being allowed.

11 FIRE PROTECTION & PREVENTION PLAN

Purpose

The Fire Protection and Prevention Plan has been developed to work in conjunction with employee emergency plans and other safety programs. This includes reviewing all new building construction and renovations to ensure compliance with applicable state, local, and national fire and life safety standards. Fire prevention measures reduce the incidence of fires by eliminating opportunities for ignition of flammable materials.

Responsibilities

Management

- Ensure all fire prevention methods are established and enforced.
- Ensure fire suppression systems such as sprinklers and extinguishers are periodically inspected and kept to a high degree of working order.
- Train supervisors to use fire extinguishers for incipient fires.
- Train employees on evacuation routes and procedures.

Supervisors

- Closely monitor the use of flammable materials and liquids.

- Train assigned employees in the safe storage, use, and handling of flammable materials.
- Ensure flammable material storage areas are properly maintained.

Employees

- Use, store, and transfer flammable materials in accordance with supplied training.
- Do not mix flammable materials.
- Immediately report violations of the Fire Protection and Prevention Plan.

Hazards

Fire and explosion hazards can exist in almost any work area. Potential hazards include:

- Improper operation or maintenance of gas fired equipment.
- Improper storage or use of flammable liquids.
- Smoking in prohibited areas.
- Accumulation of trash.
- Unauthorized hot work operations.

Hazard Control

Elimination of Ignition Sources

All nonessential ignition sources must be eliminated where flammable liquids are used or stored. The following is a list of some of the more common potential ignition sources:

- Open flames, such as cutting and welding torches, furnaces, matches, and heaters-these sources should be kept away from flammable liquids operations. Cutting or welding on flammable liquids equipment should not be performed unless the equipment has been properly emptied and purged with a neutral gas such as nitrogen.
- Chemical sources of ignition such as motors and circuit breakers-these sources should be eliminated where flammable liquids are handled or stored. Only approved explosion-proof devices should be used in these areas.
- Mechanical sparks can be produced because of friction. Only non-sparking tools should be used in areas where flammable liquids are stored or handled.
- Static sparks can be generated because of electron transfer between two contacting surfaces. The electrons can discharge in a small volume, raising the temperature to above the ignition temperature. Every effort should be made to eliminate the possibility of static sparks. Also, proper bonding and grounding

procedures must be followed when flammable liquids are transferred or transported.

Removal of Incompatibles

Materials that can contribute to a flammable liquid fire should not be stored with flammable liquids. Examples are oxidizers and organic peroxides, which, on decomposition, can generate substantial amounts of oxygen.

Control of Flammable Gases

Generally, flammable gases pose the same type of fire hazards as flammable liquids and their vapors. Many of the safeguards for flammable liquids also apply to flammable gases, other properties such as toxicity, reactivity, and corrosively also must be considered. Also, a gas that is flammable could produce toxic combustion products.

Fire Extinguishers

A portable fire extinguisher is a "first aid" device and is highly effective when used while the fire is small. The use of fire extinguisher that matches the class of fire, by a person who is well trained, can save both lives and property. Portable fire extinguishers must be installed in workplaces regardless of other firefighting measures. The successful performance of a fire extinguisher in a fire situation largely depends on its proper selection, inspection, maintenance, and distribution.

Classification of Fires and Selection of Extinguishers

Fires are classified into four general categories depending on the type of material or fuel involved. The type of fire determines the type of extinguisher that should be used to extinguish it.

Class A fires involve materials such as wood, paper, and cloth which produce glowing embers or char.

Class B fires involve flammable gases, liquids, and greases, including gasoline and most hydrocarbon liquids which must be vaporized for combustion to occur.

Class C fires involve fires in live electrical equipment or in materials near electrically powered equipment.

Class D fires involve combustible metals, such as magnesium, zirconium, potassium, and sodium.

Extinguishers will be selected according to the potential fire hazard, the construction and occupancy of facilities, hazard to be protected, and other factors pertinent to the situation.

Emergency Exits

Every exit will be clearly visible, or the route to it conspicuously identified in such a manner that every occupant of the building will readily know the direction of escape from any point. At no time will exits be blocked.

Any doorway or passageway which is not an exit or access to an exit, but which may be mistaken for an exit will be identified by a sign reading "NOT AN EXIT" or a sign indicating it actual use (i.e., "Storeroom"). Exits and accesses to exits will be marked by a readily visible sign. Each exit sign (other than internally illuminated signs) will be illuminated by a reliable light source providing not less than 5 foot-candles on the illuminated surface.

Emergency Plan for Persons with Disabilities

The supervisor is assigned the responsibility to assist persons with disabilities (PWD) under their supervision. An alternate assistant will be chosen by the supervisor. The role of the two assistants is to report to their assigned person, and to either assist in evacuation or ensure that the PWD is removed from danger.

Supervisors, alternates, and the person with a disability will be trained on available escape routes and methods. A list of PWDs is kept in the central office at each location. Visitors who have disabilities will be assisted in a manner like that of company employees. The host of the PWD will assist in their evacuation.

Emergencies Involving Fire

Fire Alarms

In the event of a fire emergency, a fire alarm will sound for the building.

Evacuation Routes and Plans

Each facility shall have an emergency evacuation plan. All emergency exits shall conform to NFPA standards. Should evacuation be necessary, go to the nearest exit or stairway and proceed to an area of refuge outside the building. Most stairways are fire resistant and present barriers to smoke if the doors are kept closed. Do not use elevators. Should the fire involve the control panel of the elevator or the electrical system of the building, power in the building may be cut and you could be trapped between floors. Also, the elevator shaft can become a flue, lending itself to the passage and accumulation of hot gases and smoke generated by the fire.

12 FLAMMABLE LIQUIDS

Purpose

Proper storage and use of flammable liquids can significantly reduce the possibility of accidental fires and injury to employees. To minimize risk to life and properly, the requirements of NFPA 30 and 321 and OSHA Standard 1910.106 have been implemented. SDSs for flammable liquids are maintained at each location.

Responsibilities

Management

- Provide proper storage for flammable liquids.
- Ensure proper training is provided to employees who work with flammable liquids.
- Ensure containers are properly labeled.

Supervisors

- Provide adequate training in the use and storage of flammable liquids.
- Monitor for proper use and storage.
- Keep only the minimum amount required on hand.
- Ensure SDS are current for all flammable liquids.

Employees

- Follow all storage and use requirements.
- Report deficiencies in storage and use to supervisors.

- Immediately report spills to supervisors.

Hazard Control

Engineering Controls

- Properly designed flammable storage areas.
- Ventilated storage areas.
- Grounding straps on drums and dispensing points.

Administrative Controls

- Designated storage areas.
- Limiting amount of flammable liquids in use and storage.
- Employee training.
- Limited and controlled access to bulk storage areas.
- Posted "Danger", "Warning" and "Hazard" signs.

Definitions

Flammable Liquid - a liquid with a flashpoint below 1000F

- Class IA - flashpoint below 730F and boiling point below 1000F
- Class IB - flashpoint below 730F and boiling point above 1000F
- Class IC - flash at or above 730F and below 1000F

Combustible Liquids - a liquid having a flashpoint at or above 1000 F

- Class II Combustibles - Flashpoint above 1000F and below 1400F
- Class III Combustibles - Flashpoint at or above 1400F
  - Subclass IIIA - flashpoint at or above 1400F and below 2000F
  - Subclass IIIB - flashpoint at or above 2000F

Substitution

Flammable liquids sometimes may be substituted by relatively safe materials to reduce the risk of fires. Any substituted material should be stable and nontoxic and should either be nonflammable or have a high flashpoint.

Storage and Usage of Flammable Liquids

Flammable and combustible liquids always require careful handling. The proper storage of flammable liquids within a work area is especially important to protect personnel from fire and other safety and health hazards.

- Flammable liquids shall be stored in NFPA-approved flammable storage lockers or in low value structures at least 50 feet from any other structure. Do not store other combustible materials near flammable storage areas or lockers.
- Bulk drums of flammable liquids must be grounded and bonded to containers during dispensing.
- Portable containers of gasoline or diesel are not to exceed 5 gallons.

- Safety cans used for dispensing flammable or combustible liquids shall be kept at a point of use.
- Appropriate fire extinguishers are to be mounted within 75 feet of outside areas containing flammable liquids, and within 10 feet of any inside storage area for such materials.
- Storage rooms for flammable and combustible liquids must have explosion-proof light fixtures.
- No flames, hot work, or smoking will be permitted in flammable or combustible liquid storage areas.
- The maximum amount of flammable liquids that may be stored in a building are:
  o 20 gallons of Class IA liquids in containers.
  o 100 gallons of Class IB, IC, II, or III liquids in containers; or
  o 500 gallons of Class IB, IC, II, or III liquids in a single portable tank.
- Flammable liquid transfer areas are to be separated from other operations by distance or by construction having proper fire resistance.
- When not in use flammable liquids shall be kept in covered containers.
- Class I liquids may be used only where there are no open flames or other sources of ignition within the possible path of vapor travel.
- Flammable or combustible liquids shall be drawn from or transferred into vessels, containers, or portable tanks within a building only through a closed piping system,

from safety cans, by means of a device drawing through the top, or from a container or portable tanks by gravity through an approved self-closing valve.

- Maintenance and operating practices shall be in accordance with established procedures which will tend to control leakage and prevent the accidental escape of flammable or combustible liquids. Spills shall be cleaned up promptly.
- Combustible waste material and residues in a building or unit operating area shall be kept to a minimum, stored in covered metal receptacles, and disposed of daily.
- Rooms in which flammable or combustible liquids are stored or handled by pumps shall have exit facilities arranged to prevent occupants from being trapped in the event of fire.
- Inside areas in which Class I liquids are stored or handled shall be heated only by means not constituting a source of ignition.

Cabinets

Not more than 120 gallons of Class I, Class II, and Class IIIA liquids may be stored in a storage cabinet. Of this total, not more than 60 gallons may be Class I and II liquids. Not more than three such cabinets (120 gallons each) may be in a single fire area except in an industrial area.

Containers

The capacity of flammable and combustible liquid containers will be in accordance with the OSHA standards.

Container Type	Flammable Liquids		Combustible Liquids		
Container	1A	1B	1C	II	III
Glass or approved plastic1	1 pt2	1 qt2	1 gal	1 gal	1 gal
Metal (other than DOT drums)	1 gal	5 gal	5 gal	5 gal	5 gal
Safety cans	2 gal	5 gal	5 gal	5 gal	5 gal
Metal drums (DOT specification)	60 gal	60 gal	60 gal	60 gal	60 gal
Approved portable tanks	660 gal	660 gal	660 gal	660 gal	660 gal

Maximum allowable capacity of containers and portable tanks

Storage Inside Buildings

Where approved storage cabinets or rooms are not provided, inside storage will comply with the following basic conditions:

The storage of any flammable or combustible liquid shall not physically obstruct a means of egress from the building or area.

- Containers of flammable or combustible liquids will remain tightly sealed except when transferred, poured, or applied. Remove only that portion of liquid in the storage container required to accomplish a particular job.
- If a flammable and combustible liquid storage building is used, it will be a one-story building devoted principally to the handling and storing of flammable or combustible liquids. The building will have two-hour fire-rated exterior walls having no opening within ten feet of such storage.
- Flammable paints, oils, and varnishes in one or five-gallon containers used for building maintenance purposes may be stored temporarily in closed containers outside approved storage cabinets or room if kept at the job site for less than ten calendar days.

Ventilation

Every inside storage room will be provided with a continuous mechanical exhaust ventilation system. To prevent the accumulation of vapors, the location of both the makeup and exhaust air openings will be arranged to provide, as far as practical, air movement directly to the exterior of the building and if ducts are used, they will not be used for any other purpose.

13 HAZARD COMMUNICATION

To ensure chemical safety in the workplace, information about the identities and hazards of the chemicals must be available and understandable to workers. OSHA's Hazard Communication Standard (HCS) requires the development and dissemination of such information:

Chemical manufacturers and importers are required to evaluate the hazards of the chemicals they produce or import and prepare labels and safety data sheets to convey the hazard information to their downstream customers.

Chemical Inventory

This organization maintains an inventory of all known chemicals in use on the job site. A chemical inventory list is available from the safety coordinator.

Hazardous chemicals brought onto the work site by this organization and/or its contractors will be included on the hazardous chemical inventory list.

Container Labeling

Chemical manufacturers and importers will be required to provide a label that includes a harmonized signal word, pictogram, and hazard statement for each hazard class and category. Precautionary statements must also be provided.

All chemicals on site will be sorted in their original or approved containers with a proper label attached, except small quantities for immediate use. Any container not properly labeled should be given to the safety coordinator for labeling or proper disposal.

Workers may dispense chemicals from original containers only in small quantities intended for immediate use. Any chemical left, after work is completed, must be returned to the original container or given to the general manager for proper handling.

No unmarked containers of any size are to be left in the work area unattended. This organization will rely on manufacturer applied labels whenever possible and will ensure that these labels are maintained.

Containers that are not labeled or on which the manufacturers label have been removed will be relabeled. This organization will ensure that each container is labeled with the identity of the hazardous chemical contained within and any appropriate hazard warnings.

SDS Availability

Employees working with a hazardous chemical may request a copy of the SDS.

Employee Training

Employees will be trained to work safely with hazardous chemicals. Employee training will include the following.

- Employers are required to train workers on the new GHS labeling system and new safety data sheets format to facilitate recognition and understanding.
- Methods that may be used to detect a release of hazardous chemical(s) in the workplace.
- Physical and health hazards associated with chemicals.
- Protective measures to be taken.
- Safe work practices, emergency responses and use of personnel protective equipment.
- Information of the Hazard Communication Standard including:
  - o Labeling and warning systems.
  - o An explanation of Safety Data Sheet.
- Management shall have all employees to complete Hazard Communication Training to ensure they understand the SDSs.

PPE Requirements

Any employee found in violation of PPE requirements as noted on the chemical SDS may be subject to disciplinary actions up to and including discharge.

Emergency Response

Any incident of overexposure or spill of a hazardous chemical/substance must be reported immediately. The general manager of this organization or his/her designee will be responsible for ensuring that proper emergency response actions are taken in leak/spill instances.

Hazards of Non-Routine Tasks

The general manager of this organization or his/her designee will inform employees of any special tasks that may arise which would involve possible exposure to hazardous materials.

Review of safe work procedure and use of required PPE will be conducted prior to the start of such tasks. Where necessary, areas will be posted to indicate the nature of the hazard involved.

Informing Other Employees

Other on-site employees are required to adhere to the provisions of the Hazard Communication Standard.

Information on hazardous chemicals known to be present will be exchanged with other employees. Employers will be responsible for providing necessary information to the employees.

Other on-site employees will be provided with a copy of his organization's Hazard Communication Program.

14 HAZWOPER

Purpose and Scope

This organization is committed to providing a safe and healthy work environment and to protecting our employees from injury or death caused by uncontrolled hazards in the workplace. The purpose of the HAZWOPER program is to establish work policies, practices, and procedures that employees are to follow during an emergency response to a hazardous substance release/spill and during post-emergency operations. In this program, hazardous substance is defined as a substance in solid, liquid, or gaseous form that can harm humans, other living organisms, or the environment.

It will cover all areas where employees may be exposed to substances that can result in adverse health and safety effects (e.g., ammonia, Freon, gasoline, diesel fuel, battery acid and water treatment chemicals). This program is integrated into our company's written safety and health program and is a collaborative effort that includes all employees.

Hazwoper Program Responsibilities

Program Administrator.

The Program Administrator is responsible for directing all hazardous waste site operations, the HAZWOPER program implementation, management, and recordkeeping requirements.

Incident Commander.

The Incident Commander is responsible for managing emergency activities at a hazardous release site and directs the activities through a chain of command to those responsible for carrying out a specific emergency response task. The Incident Commander will also:

- Identify hazardous substances at the site.
- Enforce the incident command system procedures.
- Ensure those responding wear appropriate PPE.
- Keep others away from the site.
- Commence appropriate decontamination procedures after the emergency.

Pre-emergency Planning and Incident Command System

Pre-emergency planning and coordination with outside parties is the heart of this organization's HAZWOPER Program. All hazardous materials are identified within the Hazard Communications Program and this information is shared with the local Fire Department and Emergency Management Coordinator annually. This coordination allows the local emergency responders to have knowledge of what hazardous materials are present in our facilities and how our employees are going to react to a release or spill.

Our Incident Command System is a standardized incident management system based on federal and state models. It is designed specifically to allow responders to understand the organizational structure local emergency responders will be using on-site.

Employee Training

Those who will or may respond to an emergency will be appropriately trained before participating in an actual incident. New employees will be trained upon hiring and re-trained any time the employee's responsibilities under the plan change or whenever the plan itself changes. This organization will provide copies of all emergency response plans to employees and copies will be posted for employees to review.

First Responder Awareness Level Training.

All employees are required to receive on-site First Responder Awareness Level training. Employees will receive training prior to initial assignment and refresher training annually. All training records will be maintained on the training record.

Employees will be trained on:

- Incident Command System
- Emergency Response Plan
- Hazard Communication Program

- Areas where hazardous substances may be present
- Methods and observations that can be used to detect the presence or release of a hazardous substances in the work area
- Protective measures, including training in proper selection and use of PPE
- An explanation of the chemical labeling system
- Notification of appropriate personnel
- The elements of the Confined Space Program (if applicable)
- The medical surveillance programs.

First Responder Operations Level Training.

All supervisors and anyone at who responds to the release of a hazardous substance or contains the release but is not involved in stopping the release is required to receive on-site First Responder Operations Level training. The training includes all the topics covered in the First Responder Awareness Level training and the proper procedures for these selected employees to take if they witness, discover or otherwise become aware of a release of hazardous substances. Initial training and competencies will take approximately eight hours and employees will receive refresher training annually.

Incident Commander.

Any emergency responder expected to perform as an Incident Commander should be trained to fulfill the obligations of the position. Incident Commander training will include all the topics covered in First Responder Awareness, First Responder Operations level training and training on:

- Analyzing a hazardous substance to determine the magnitude of the release.
- Planning and implementing an appropriate response
- Evaluating the progress of the emergency response.

Initial training and competencies will take approximately 24 hours. Incident Commanders will receive refresher training annually.

Lines of Communication

In the event an employee (First Responder Awareness Level) witnesses or discovers a release of a hazardous substance, he/she will:

- Notify the Incident Commander of the release and/or call 911 if necessary.
- Follow instructions from the Incident Commander and/or supervisors.

In the event a supervisor (First Responder Operations Level) discovers or otherwise becomes aware of a release of a hazardous substance, he/she will:

- Notify Incident Commander of the release and/or call 911 if necessary.
- Communicate spill to employees and facilitate evacuation/relocation procedures.
- Assist the Incident Commander, as necessary.
- Wait for emergency response.

In the event of an emergency, the Incident Commander will:

- If necessary, call 911 and notify the proper authorities of the release of the material.
- Notify all supervisors of release and instruct them to evacuate/relocate.
- Evacuate the immediate area and keep others from entering the area.
- Identify materials from shipping or container labels (if possible, without entering the area).
- Identify proper PPE from the safety data sheets or labels.
- Determine if spill response measures can be done safely with available PPE.
- If the release can be safely contained, obtain proper material such as absorbent materials from the spill response kits on site.

- Obtain and put on needed PPE.
- Apply absorbent material or other containment measures on and around the spill or release.
- Keep other employees out of the release area.
- Wait for emergency response.

Evacuation Procedures, Safe Distances and Places of Refuge

During an emergency evacuation, employees will evacuate using specified routes to designated assembly areas. Once at the assembly area, supervisors will take role and note any missing employees and report this information to the Incident Commander or emergency responders.

Emergency Alerting and Response Procedures

This organization has an alarm system in place to inform all affected employees of a release or potential release of a hazardous substance. If the alarm is sound, affected employees should respond accordingly.

Alarm Type: Response:

Alarm Type: Response:

Alarm Type: Response:

Site Safety and Control

If the spill/release requires a full evacuation of the building, First Responders Operations Level employees will ensure no unauthorized person enters the building and wait for emergency response to arrive and secure the site.

If the spill/release does not require a full evacuation, the Incident Commander will block off the spill or release area and make sure all other employees are a safe distance away.

Emergency Medical Treatment and First Aid

In the event the spill/release causes injury to an employee requiring emergency medical treatment or first aid, the injured employee(s) will be sent to:

Hospital Name:

Address:

Phone Number:

This hospital has been made aware of our HAZWOPER Program and will handle decontamination efforts if necessary.

Medical Surveillance

- This organization has a medical surveillance program for all employees whose role may expose them to hazardous substances at or above the permissible exposure limits (PEL) or, if there is no PEL, above the published exposure levels for more than 30 days per year.

- All employees whose job includes First Responder Operations Level duties or Incident Commander duties will be medically examined before assignment, immediately after reporting symptoms of possible overexposure and at termination or reassignment from the First Responder Operations Level or Incident Commander duties. Additionally, all employees who are or may be exposed to a hazardous substance or who are required to wear a respirator for 30 days or more a year are covered under this medical surveillance program.

- All medical surveillance examinations are performed by licensed physicians, without cost to the employee and at a reasonable time and place.

- Examinations will include a medical and work history. Special emphasis will be placed on symptoms of exposure to hazardous substances and health hazards on the job and to fitness for duty, including the ability to wear PPE during their First Responder Operations Level or Incident Commander role.

- This organization has provided the examining physician a copy of the HAZWOPER Program and its appendices, a description of the employee's duties, anticipated exposure levels, a description of any PPE to be used, and any information from previous medical examinations.

- This organization will obtain a written opinion from the physician that contains the results of the medical examination, any detected medical conditions that could place the employee at an increased risk of exposure, and any recommended limitations.
- The employee will receive a copy of the physician's written findings. The written opinion obtained by the employer shall not reveal specific findings or diagnoses unrelated to the possible occupational exposure.
- These medical opinions and other related information will be kept in the employee's HR personnel file.
- All exposure records, medical examination records, and written opinions will be maintained for the duration of employment plus 30 years.
- Each First Responder Operations Level employee and Incident Commander will be notified in writing that their medical surveillance records are maintained in the Human Recourses department in their personnel file and are available during regular business hours.

Chemical-Protective Equipment

- PPE will be available in the spill response kits and will also be provided to each First Responder Operations Level employee and Incident Commander.
- Employees will be trained on how to safely inspect, use and wear the PPE listed below.
- Employees should only perform their spill response duties if the provided PPE will allow safe interaction

with the released chemicals as per the safety data sheets.

- The PPE distributed will be determined in the Personal Protective Equipment Program assessment and based on the exposures determined during that assessment.
- PPE will be updated based on this assessment at least annually.

Post-Emergency Response Operations (Decontamination)

This organization's maintenance employees will perform spill clean-up operations if the amount of material released is at an incidental level. If the release is of an amount beyond incidental, outside services will be used.

Periodic Program Review

The Hazwoper Program and procedures will be reviewed annually.

15 HEARING PROTECTION

Purpose

To comply with governmental requirements 29 CFR 1910 and to prevent noise induced hearing loss.

Affected Employees

Those employees who are subjected to noise levels exceeding those listed in the following table.

Methods Of Protection

- Engineering Controls: Whenever feasible engineering shall be the first choice in defending against noise levels exceeding the permissible exposure limit (PEL). This may be accomplished by redesigning the equipment or process or just by simply adding a muffler to an exhaust system.
- Administrative Controls: Administrative controls are simply reducing the amount of time someone is exposed to the hazard. Therefore, if noise levels cannot be reduced, the amount of time the employee is exposed shall be lowered to meet the requirements.

Personal Protective Equipment:

Whenever it is not feasible to use either engineering or administrative controls, personal protective equipment shall be made available.

The use of personal protective equipment to prevent hearing loss is mandatory when levels exceed the table noted above (Permissible Noise Levels).

When ear plugs are used for protection, they must be properly fitted to provide adequate protection.

Non-disposable plugs shall be cleaned after each use.

When earmuff type protection is used, it must form a perfect seal. Beards, glasses, long hair and even chewing gum can reduce the effectiveness of earmuffs.

Plain cotton shall not be used as hearing protection.

16 HEAT STRESS

Purpose

The Heat Stress Program has been developed to provide guidance and oversight for the activities involving elevated temperatures during occupational activities.

Scope

The Heat Stress Program applies to all employees, contractors and students working company facilities.

Responsibility

Heat Stress Program Manager

- Assist in the identification of elevated heat work areas.

- Ensure the proper testing, monitoring, and documentation for suspect and known elevated heat work areas.

- Ensure all equipment used for testing and monitoring is appropriate and in proper working condition.

- Document the program to manage the occupational activities in elevated heat work areas.

- Maintain an inventory of elevated heat work areas.

- Provide or ensure the training as necessary to personnel required to work elevated heat work areas.

- Assist in the development of localized administrative, engineering or PPE controls and measures to reduce or eliminate heat stress conditions.

- Conduct periodic review of the heat stress program to ensure it follows federal guidelines, regulations and best practices.

Supervisors

- Assist in the identification of elevated heat work areas.

- Monitor suspect locations for elevated heat conditions.

- Notify the Heat Stress Program Manager or representative of suspect hazardous work conditions involving elevated heat conditions.

- Implement required engineering controls, instrumentation changes, or work practice changes requested to reduce heat load.

- Maintain a copy of this written program in the workplace.

- Ensure that employees required to work under suspect elevated heat conditions are trained in the heat stress program.

Non-supervisory Employees:

- Assist in the identification of elevated heat work areas

- Attend required training(s) as specified by a supervisor or the Heat Stress Program Manager.

- Comply with procedures as required by the Heat Stress Program, and all other heat stress related guidance as deemed appropriate by a supervisor.

- Use all personal protective equipment as specified in prescribed training or required by a supervisor or the Heat Stress Program Manager.

- Immediately notify a supervisor, Heat Stress Program Manager, or Safety Officer. of any hazards encountered.

Heat Stress Safety Program

The Heat Stress Program was established to promote health and safety of occupational activities in locations where elevated temperatures and humidity exist.

Thermal heat stress includes injuries or illnesses caused when a person is working in or exposed to elevated temperature conditions directly affecting an individual's ability to function in a normal manner.

The intent of the program is to ensure a core body temperature of individuals as close to normal (typically within 1 Celsius). The program functions by limiting the amount of work time an individual is exposed to elevated temperatures while completing specific types of workloads (low, medium, high). A key element of the program involves identifying sources of heat load and planning and measures to reduce or eliminate that heat load where possible.

Monitoring

Upon notification by an employee, the Heat Stress Program Manager will:

- Work with a knowledgeable individual who can adequately describe the perceived sources of heat load and time periods that are both typical and worse case for the suspect location

- Use a WBGT instrument that is within calibration.

- Setup the WBGT in the suspect location allowing for a 15 minute "warm up" period before logging values,

- Begin documenting the readings from the instrument on Appendix A: Monitoring Sheet,

- Allow the unit to run for an appropriate period to adequately capture the heat intensity that is created by either natural environmental conditions or equipment/instrumentation that adds heat load,

- Provide technical feedback to the Supervisor or representative of the work area to allow an immediate response to ensure the safety of personnel.

Heat Related Conditions:

A. **Minor conditions:**

Heat edema presents with swelling and discomfort of the hands and feet. Individuals may complain that their shoes feel tight or are ill fitting. The exact cause is unknown but generally involves dilation of the blood vessels and shifts in fluid within the body. The condition is self-limiting, and symptoms typically resolve within a few days.

Miliaria rubra, also known as prickly heat or heat rash occurs when sweat gland pores become blocked. Sometimes a secondary infection may occur. Skin with Miliaria rubra cannot sweat effectively. Therefore, the risk of heat illness is increased in proportion to the amount of skin involved.

Sunburn impairs sweating and predisposed to heat injury from systemic effects, including fever, that influence thermoregulation. When sunburn occurs over 5% of the body surface area, the effected individual should be kept from significant heat strain until the burn has healed.

Heat tetany may result when an individual hyperventilates after being exposed to heat stress. Symptoms include muscle spasms and numbness and tingling around the mouth. It generally occurs before heat acclimatization.

Syncope is a temporary circulatory failure due to pooling of the blood in the peripheral veins. Symptoms range from lightheadedness to loss of consciousness. Victims typically recover rapidly once they sit or lay supine. Syncope occurring more than five days after heat exposure may indicate dehydration or heat exhaustion.

Heat cramps are brief, recurrent, often painful skeletal muscle cramps. The cramps are usually preceded by muscle fasciculations which may be seen or felt on the muscle surface. Cramps produce a hard lump in the muscle. There are no systemic symptoms.

Major Conditions:

<u>**Heat Exhaustion**</u> is the most common heat related cause of illness. It occurs when the heart cannot pump quickly enough to sustain the needs of the skin blood flow to maintain body temperature along with the metabolic needs of the body for muscle and vital organ activity. Dehydration, reduced blood volume and constricted blood vessels are all contributing factors. The signs and symptoms of heat exhaustion include:

- Generalized weakness
- Headache
- Nausea
- Fatigue
- Dizziness
- Increased heat rate and muscle cramps
- Sweating persists and may even be profuse, and the individual may become disoriented.

Treatment should begin immediately to prevent progression to a more severe heat injury. The more severe assessment of heat stroke should be assumed in anyone who experiences a change in mental status such as disorientation.

Heat Stroke is characterized by elevated body temperature (>104° F) and dysfunction of the central nervous system resulting in delirium, convulsions or coma. Two types of heat stroke may occur, exertional and classical.

- Exertional heat stroke occurs in physically active individuals who are producing substantial metabolic heat. It is the most usual form in workers and athletes and can occur in both hot and temperate conditions.

- Classical heat stroke occurs in vulnerable populations such as the young, the elderly and those without potable water. This type often presents as an epidemic during summer heat waves.

- For up to an hour prior to the onset of heat stroke one may experience:

  o Headache

  o Dizziness

  o Drowsiness

  o Restlessness

  o Confusion

  o Irrational or aggressive behavior

Heat measures for employees

- Administratively rotating the workload among workers
- Shortening the duration of exposure
- Increasing rest time or frequency of rest periods (rest periods should be in a cooler shaded area in relation to the work area)
- Utilizing air-conditioned rest areas
- Minimizing direct heat sources by insulation or installing reflective screening
- Using additional fans in the work area
- Providing cool drinking water

Heat Stress Safety Training

All supervisors of employees who work in elevated heat areas should ensure that heat stress training is available. Types of training include a formal presentation by the Heat Stress Program Manager, "Tool-box talks" led by the supervisor, or computer-based training. Training should incorporate both identifying heat related conditions and measure to overcome elevated heat conditions.

Medical Surveillance

Employees who routinely work under heat stress conditions should be required to have an annual physical. Factors that must be considered before assigning an employee to work under heat stress conditions include:

1. **Acclimatization**

- Adaptation to new surroundings or conditions

- Under heat stress conditions, gradually increase the work time over the first two weeks. (Individuals are most susceptible to heat stress during their initial two-week period of work in hot and humid conditions).

- Acclimatization can be affected by:

    o Medical condition or Prescription Medication, inform your supervisor prior to beginning work under heat stress conditions.

    o Allergy medicine (prescription or non-prescription

    o Sunburn – likely to reduce work capacity under heat stress conditions.

2. **Muscular activity and work capacity**
3. **Age / physical condition**
4. **Prescription drug use**

17 HOUSEKEEPING

Purpose

To provide a clean and healthy workplace and to comply with regulatory requirements.

General Requirements

The following requirements represent the minimum acceptable standard of housekeeping.

- Daily clean-up of work, maintenance, and personnel areas is required.
- All equipment and process materials shall be stored in an orderly manner.
- All designated areas utilized for temporary storage of materials shall be properly barricaded.
- All scrap materials and waste shall be picked up and disposed of. Debris shall be placed into waste containers provided.
- Stairways, walkways, and fixed ladder areas shall be kept clear of all cords, cables, hoses, materials, and anything else that might hinder personnel passage.
- Cords, cables, and hoses at stairways, walkways, and scaffolds shall be supported at least seven feet overhead or laid flat outside of walkways.
- All spills of oil, solvents, chemicals, and any regulated liquids shall be reported immediately. Dumping of these materials into floor drains, sanitary sewers, storm

sewers, drainage ditches, or other open ground is forbidden.

- Loose materials on roofs or other overhead structures shall be removed or secured to prevent being blown or bumped off.
- Accumulation of materials that may create a fire hazard shall not be permitted.

18 LADDER SAFETY

Purpose

The purpose of this safety policy and procedure is to establish guidelines for the safe use of ladders throughout the facility by employees and contractors.

Ladders are used when employees need to move up or down between two different levels. Slips, trips, and falls can be significant contributors to accidents. Slips, trips, and falls can occur when wrong ladder selection is made and when improper climbing techniques and/or defective ladders are used.

The appropriate ladder will be used for the corresponding job and defective ladders will not be used. When hazards exist that cannot be eliminated, then engineering practices, administrative practices, safe work practices, PPE, and proper training regarding ladders will be implemented. These measures will be implemented to minimize those hazards to ensure the safety of employees and the public.

Responsibilities

Managers

Managers are responsible for ensuring that adequate funds are available and budgeted for the purchase of ladders in their areas. Managers will obtain and coordinate the required training for the affected employees. Managers will also ensure compliance with this safety policy and procedure through their auditing process.

Supervisors

Supervisors are responsible for ensuring that all ladders (fixed and portable) are regularly inspected and properly maintained. They will also be responsible for tagging ladders in need of repair and removing defected ladders from service for repair or destruction. Supervisors will audit for compliance with this safety policy and procedure during their facility and jobsite audits.

Employees

Employees shall comply with all applicable guidelines contained in this safety policy and procedure. Employees are also responsible for reporting immediately suspected unsafe conditions or ladders to their supervisor. Employees are to inspect ladders before using and keep ladders clean and in good condition.

Training

Employees using the ladders shall be trained in:

- The proper use of the ladders.
- What kind of ladder to use.
- How to set up ladders.
- Ladder inspection; and
- Proper maintenance.

This training shall be done upon initial employment and/or job assignment. Refresher training shall be provided to employees at the discretion of their supervisor.

Ladder Hazards and Safe Use

Ladder Hazards

There are inherent hazards associated with ladder use. Typical ladder hazards include:

- Insufficient surface resistance on ladder rungs and steps.
- Ladder structural failure.
- Ladders tipping sideways, backwards, or slipping out at the bottom.
- Ladder spreaders not fully opened and locked, causing the ladder to "walk", twist or close when a load is applied to the ladder.
- Using metal ladders around electricity.
- Using deteriorated ladders; and
- Using fixed ladders without cages or fall protection.

Safe Ladder Use

- Hold on with both hands when going up or down. If material must be handled, raise or lower it with a rope either before going down or after climbing to the desired level.
- Always face the ladder when ascending or descending.
- Never slide down a ladder.

- Be sure shoes are not greasy, muddy, or slippery before climbing.
- Do not climb higher than the third rung from the top on straight or extension ladder, or the second tread from the top on stepladders.
- Carry tools on a tool belt not in the hand.
- Never lean too far to the sides. Keep your belt buckle within the side rails.
- Use a 4 to 1 ratio when leaning a single or extension ladder (place a 12-foot ladder so that the bottom is 3 feet away from the object the ladder is leaning against).
- Inspect ladder for defects before using.
- Never use a defective ladder. Tag or mark it so that it will be repaired or destroyed.
- Never splice or lash a short ladder together.
- Never use makeshift ladders, such as cleats fastened across a single rail.
- Be sure that a stepladder is fully open, and the metal spreader locked before starting to climb.
- Keep ladders clean and free from dirt and grease.
- Never use ladders during a strong wind except in an emergency and then only when they are securely fastened.
- Never leave placed ladders unattended.
- Never use ladders as guys, braces, or skids, or for any other purpose other than their intended purposes.
- Never attempt to adjust a ladder while a user is standing on the ladder.

- Never jump from a ladder. Always dismount from the bottom rung.

Ladder Inspection

An inspection program should be set up by which all ladders are inspected once every three months. Ladders that are weak, improperly repaired, damaged, have missing rungs, or appear unsafe shall be removed from the job or site for repair or disposal. Before discarding a wood ladder, cut it up so no one can use it again. Additionally, portable ladders must be always maintained in good condition and inspected frequently. Tag any ladders that have developed defects with "DANGEROUS - DO NOT USE" and remove from service for repair or disposal.

For portable wood ladders, all wood parts shall be free from sharp edges and splinters; sound and free from accepted visual inspection from shake, wane, compression failures, decay, or other irregularities. For portable metal ladders, the design shall be without structural defects or accident hazards such as sharp edges, burrs, etc. The selected metal shall be of sufficient strength to meet the test requirements and shall be protected against corrosion. For fixed ladders, all wood parts shall meet the criteria of wood ladders. All metal parts shall meet the criteria of metal ladders.

Maintenance

Portable wood ladders may be coated with a water-repellent preservative to provide a suitable protective material. Metal ladders and metal parts on wood ladders should be corrosion-resistant and kept free from nicks. If nicks occur, they should be promptly treated to prevent possible metal fatigue due to rust.

Ladder Inspection Checklist

All Ladders

- Loose steps or rungs (if they can be moved at all with the hand).
- Loose nails, screws, bolts, or other metal parts.
- Cracked, split, or broken uprights, braces, steps, or rungs.
- Slivers on uprights, rungs, or steps.
- Damaged or worn non-slip bases.
- Rusted or corroded spots.

Stepladders

- Wobbly from side strain.
- Loose or bent hinge spreaders.
- Stop on hinge spreaders broken.
- Broken, split, or worn steps; and
- Loose hinges.

Extension Ladders

- Loose, broken, or missing extension locks.
- Defective locks that do not seat properly when the ladder is extended; and
- Deterioration of rope, from exposure to weather, acid or other destructive agents.

Fixed Ladders

- Loose, worn, or damaged rungs or side rails.
- Damaged or corroded parts of cage.
- Corroded bolts and rivet heads on inside of metal stacks.
- Damaged or corroded handrails or brackets on platforms.
- Weakened or damaged rungs on brick or concrete slabs; and
- Base of ladder obstructed.

19 LOCKOUT/TAGOUT

Purpose

This procedure establishes guidelines to protect the safety of all personnel involved in maintenance of servicing operations. It shall be used to ensure that the machine or equipment is locked out or tagged out before any servicing or maintenance activities are performed where the unexpected energization, start-up, or release of energy could cause injury.

Definitions

Affected Employee:

An employee whose job requires him/her to operate or use a machine or equipment on which servicing, or maintenance is being performed under lockout or tagout, or whose job requires him/her to work in an area in which such servicing or maintenance is being performed.

Authorized Employee:

A person who locks or implements a tagout system procedure on machines or equipment to perform the servicing or maintenance on that machine or equipment. An authorized employee and an affected employee may be the same person when the affected employee's duties also include performing maintenance or service on a machine or equipment which must be locked or a tagout system implemented.

Capable of Being Locked Out:

An energy isolating device will be capable of being locked out either if it is designed with a hasp or other attachment or integral port to which or through which a lock can be affixed, or if it has a locking mechanism built into it.

Energized:

Connected to an energy source or containing residual or stored energy.

Energy Isolating Device:

A mechanical device that physically prevents the transmission or release of energy.

Energy Source:

Any source of electrical, mechanical, hydraulic, pneumatic, chemical, thermal, or other energy

Lockout:

The placement of a lockout device on an energy isolating device, in accordance with an established procedure, ensuring that the energy isolating device and the equipment being controlled cannot be operated until the lockout device is removed.

Lockout Device:

A device that utilizes a positive means such as a lock, either key or combination type, to hold an energy isolating device in the safe position and prevent the energizing of a machine or equipment.

Tagout:

The placement of a tagout device on an energy isolating device, in accordance with an established procedure, to indicate that the energy isolating device and equipment being controlled may not be operated until the tagout device is removed.

Tagout Device:

A prominent warning device, such as a tag and a means of attachment, which can be securely fastened to an energy isolating device in accordance with an established procedure, to indicate that the energy isolating device and the equipment being controlled may not be operated until the tag is removed.

Responsibility and Training

Appropriate employees shall be instructed in the safety significance of the lockout (or tagout) procedure. All employees involved in maintenance or servicing operations shall be authorized to lockout or tagout. Each affected employee or employees whose work operations are in the area of a lockout/tagout, shall be instructed in the purpose and use of the lockout or tagout procedure.

When servicing and/or maintenance is performed by a crew, craft, department, or other group, group lockout or tagout devices shall be used in accordance with the requirements set forth in this procedure. Primary responsibility shall be vested in an authorized employee for a set number of employees working under the protection of a group lockout or tagout device.

General Procedures

Make a survey to locate and identify all isolating devices to be certain which switch(s), valve(s), or other energy isolating devices apply to the equipment to be locked or tagged out. More than one energy source (electrical, mechanical or others) may be involved.

Notify all affected employees that a lockout or tagout system is going to be utilized and the reason, therefore. The authorized employee shall know the type and magnitude of energy that the machine or equipment utilizes and shall be knowledgeable of the hazards thereof.

If the machine or equipment is operating, shut it down by the normal stopping procedures (depress stop button, turnkey to off position, open toggle switch, etc.).

Stored energy (such as that in springs, elevated machine members, rotating flywheels, hydraulic systems, and air gas, steam, or water pressure, etc.) must be dissipated or restrained by methods such as repositioning, blocking, bleeding down, etc.

The authorized employee and/or affected employee shall lockout and/or tagout the energy isolating device.

After ensuring that no personnel are exposed, operate the push button or other normal operating controls to make certain the equipment will not operate. Return operating controls to "neutral" or "off" position after the test.

Each affected employee who has placed a lockout lock on an energy isolating device shall be the only custodian of his/her key. Where group lockouts are used, the authorized employee shall be the custodian of the group lockout key.

If the equipment does not start when it is tried, then all energy source and lockouts must be rechecked. No work may be performed on the equipment until a successful lockout has been confirmed.

The equipment is now locked out or tagged out. Where lockout tags are used the tag shall be labeled "DANGER - DO NOT OPERATE" and must be signed and dated by the authorized and/or affected employee.

In the event of a shift change while servicing or maintenance is being performed under the lockout/tagout procedure, the authorized or affected employee(s) shall transfer his/her personal lock(s), key(s) to the oncoming employee who is relieving him/her. This transfer will be recorded in the lockout/tagout record book. When he/she received the lockout key(s), the oncoming employee becomes the authorized and/or affected employee. If locks are not used and equipment is under a tagout procedure, tags shall remain in place and an appropriate entry in the record book shall be made.

Lockout Removal Procedures

After the servicing and/or maintenance is complete and equipment is ready for normal production operations, check the area around the machine or equipment to ensure that none is exposed.

After all tools have been removed from the machine or equipment, guards have been reinstalled and employees are in the clear, remove all lockout or tagout devices. Operate the energy isolating device to restore energy to the machine or equipment.

Procedure Involving More Than One Person

In the preceding steps, if more than one individual is required to lockout or tagout equipment, each shall place his/her own personal lockout device or tagout device assigned by the company on the energy isolating devices(s).

When an energy isolating device cannot accept multiple locks or tags, a multiple lockout or tagout device (hasp) may be used. If lockout is used, a single lock may be used to lockout the machine or equipment with the key being placed in a lockout box or cabinet. As each person no longer needs to maintain his or her lockout protection, that person will remove his/her lock from the box or cabinet.

Temporarily Energizing for Testing and Positioning

There may also be cases where a piece of equipment that has been serviced under lockout or tagout must be temporarily energized for testing, positioning, etc. In these situations, the authorized employee shall notify all affected employees to clear the area prior to removal of the lockout and startup of the equipment or machine.

The procedures identified for removal of equipment from lockout shall be employed. After testing is complete, equipment shall again be deenergized and locked out in accordance with this procedure and shall be reinstated before servicing is resumed.

Basic Rules for Using Lockout or Tagout Procedures

- All equipment shall be locked out or tagged out to protect against accidental or inadvertent operation when such operation could cause injury to personnel.
- Never attempt to operate any switch, valve, or other energy isolating device where it is locked or tagged out.
- Never attempt to short cut the lockout/tagout procedure. This may endanger your personal safety as well as others.
- If an affected or authorized employee leaves a lock in place and that lock needs to be removed after they have left the workplace:
  o Every reasonable attempt will be made to contact them.

- o If they cannot be reached, the authorized employee and his supervisor will cut the lock off.
- o The appropriate logbook entry will be made; and
- o The employee will be notified when they return to work that their lock was removed.

20 MACHINE SAFETY

Purpose

The Machine Guard Program is designed to protect Employees from hazards of moving machinery. All hazardous areas of a machine shall be guarded to prevent accidental "caught in" situations.

Responsibilities

Management

- Ensure all machinery is properly guarded
- Provide training to employees on machine guard rules
- Ensure new purchased equipment meets the machine guard requirements prior to use

Supervisors

- Train assigned employees on the specific machine guard rules in their areas
- Monitor and inspect to ensure machine guards remain in place and functional
- Immediately correct machine guard deficiencies

Employees

- Do not remove machine guards unless equipment is locked and tagged
- Replace machine guards properly
- Report machine guard problems to supervisors immediately

- Do not operate equipment unless guards are in place and functional.
- Only trained and authorized employees may remove machine guards.

Definition of Terms

- Guards: Barriers that prevent Employees from contact with moving portions or parts of exposed machinery or equipment which could cause physical harm to the Employees.
- Enclosures: Mounted physical barriers which prevent access to moving parts of machinery or equipment.
- Point-of-Operation: The area on a machine or item of equipment, where work is being done and material is positioned for processing or change by the machine.
- Power Transmission: Any mechanical parts which transmit energy and motion from a power source to the point-of-operation. Example: Gear and chain drives, cams, shafts, belt and pulley drives and rods. NOTE: Components which are (7) feet or less from the floor or working platform shall be guarded.
- Nip Points: In-Running Machine or equipment parts, which rotate towards each other, or where one part rotates toward a stationery object.
- Shear points: The reciprocal (back and forth) movement of a mechanical part past a fixed point on a machine.
- Rotating Motions of an exposed mechanism are dangerous unless guarded. Even a smooth, slowly

rotating shaft or coupling can grasp clothing or hair upon contact with the skin and force an arm or hand into a dangerous position. Affixed or hinged guard enclosure protects against this exposure.

- Reciprocating: Reciprocating motions are produced by the back-and-forth movements of certain machine or equipment parts. This motion is hazardous, when exposed, offering pinch or shear points to an Employee. A fixed enclosure such as a barrier guard is an effective method against this exposure.

- Transverse Motions: Transverse motions are hazardous due to straight line action and in-running nip points. Pinch and shear points also are created with exposed machinery and equipment parts operating between a fixed or other moving object. A fixed or hinged guard enclosure provides protection against this exposure.

- Cutting Actions: Cutting action results when rotating, reciprocating, or transverse motion is imparted to a tool so that material being removed is in the form of chips. Exposed points of operation must be guarded to protect the operator from contact with cutting hazards, being caught between the operating parts and from flying particles and sparks.

- Shearing Action: The danger of this type of action lies at the point of operation where materials are actually inserted, maintained and withdrawn. Guarding is accomplished through fixed barriers, interlocks, remote control placement (2 hand controls), feeding or ejection.

Hazards

Use of machinery or equipment with inadequate guards or damaged controls can result in:

- Amputation
- Skin Burns
- Cuts & fractures
- Death

Hazard Controls - controls used to prevent exposure to moving or energized machine parts includes:

- Machine guards
- Interlocks
- Presence sensing devices
- Gates
- Two-hand controls
- Employee training

Machine Guarding Requirements

- Guards shall be affixed to the machine where possible and secured.
- A guard shall not offer an accident hazard in itself.
- The point-of-operation of machines whose operation exposes an Employee to injury shall be guarded.
- Revolving drums, barrels and containers shall be guarded by an enclosure which is interlocked with the drive mechanism.
- When periphery of fan blades are less than 7 feet above the floor or working level the blades shall be guarded with a guard having openings no larger than 1/2 inch.
- Machines designed for a fixed location shall be securely anchored to prevent walking or moving. For example, Drill Presses, Bench Grinders, etc.

General Requirements for Machine Guards

- Guards must prevent hands, arms, or any part of an employee's body from making contact with hazardous moving parts. A good safeguarding system eliminates the possibility of the operator or other Employees from placing parts of their bodies near hazardous moving parts.
- Employees should not be able to easily remove or tamper with guards. Guards and safety devices should be made of durable material that will withstand the conditions of normal use and must be firmly secured to the machine.

- Guard should ensure that no objects can fall into moving parts. An example would be a small tool which is dropped into a cycling machine could easily become a projectile that could injure others.
- Guard edges should be rolled or bolted in such a way to eliminate sharp or jagged edges.
- Guard should not create interference which would hamper Employees from performing their assigned tasks quickly and comfortably.
- Lubrication points and feeds should be placed outside the guarded area to eliminate the need for guard removal.

Training

All Employees shall be provided training in the hazards of machines and the importance of proper machine guards. Machine safety and machine guarding rules will be thoroughly explained as part of the new hire orientation program and annually as refresher safety training.

21 OFFICE SAFETY

Purpose

The purpose of this program is to provide guidance to office managers and office staff on the elements of safe office work. The office is like any other work environment in that it may present potential health and safety hazards; however, most of these may be minimized or eliminated by designing jobs and workplaces properly, and by considering differences among tasks and individuals. Inadequate environmental conditions, such as noise, temperature, and humidity, may cause temporary discomforts. Environmental pollutants such as chemical vapors released from new carpeting and furniture may also induce discomforts.

Responsibilities

Management

- Provide training for all office staff in:
  - o Emergency Procedures
  - o Electrical Safety
  - o Office Ergonomics
- Ensure office equipment is in safe working order.
- Provide proper storage for office supplies.

Office Staff

- Report all safety problems immediately.

- Not attempt to repair any office equipment or systems.

- Maintain a neat and sanitary office environment.

Electrical Safety

Electric cords should be examined on a routine basis for fraying and exposed wiring. Particular attention should be paid to connections behind furniture, since files and bookcases may be pushed tightly against electric outlets, severely bending the cord at the plug. Electrical appliances must be designed and used in accordance with UL requirements.

Use of Extension Cords

- Extension cords shall only be used in temporary situations where fixed wiring is not feasible.

- Extension cords shall be kept in good repair, free from defects in their insulation. They will not be kinked, knotted, abraded, or cut.

- Extension cords shall be placed so they do not present a tripping or slipping hazard.

- Extension cords shall not be placed through doorways having doors that can be closed, and thereby damage the cord.

- All extension cords shall be of the grounding type (three conductor).

Housekeeping

Good housekeeping is an essential element of accident prevention in offices. Poor housekeeping may lead to fires, injuries to personnel, or unhealthful working conditions. Mishaps caused by dropping heavy cartons and other related office equipment and supplies could also be a source of serious injuries to personnel.

Passageways in offices should be free and clear of obstructions. Proper layout, spacing, and arrangement of equipment, furniture, and machinery are essential. All aisles within the office should be clearly defined and kept free of obstructions.

Chairs, files, bookcases, and desks must be replaced or repaired if they become damaged. Damaged chairs can be especially hazardous.

Filing cabinet drawers should always be kept closed when not in use. Heavy files should be placed in the bottom file drawers.

Materials stored within supply rooms must be neatly stacked and readily reached by adequate aisles. Care should be taken to stack materials so they will not topple over.

Under no circumstances will materials be stacked within 18 inches of ceiling fire sprinkler heads.

Materials shall not be stored so that they project into aisles or passageways in a manner that could cause persons to trip or could hinder emergency evacuation.

Computer Workstations

Complaints concerning musculoskeletal problems are frequently heard from computer operators. The most common are complaints relating to the neck, shoulders, and back. Others concern the arms and hands and occasionally the legs.

Certain common characteristics of VDT jobs have been identified and associated with increased risk of musculoskeletal problems. These include:

- Design of the workstation.

- Nature of the task.

- Repetitiveness of the job.

- Degree of postural constraint.

- Workspace.

- Work/rest schedules; and

- Personal attributes of individual workers.

The key to comfort is in maintaining the body in a relaxed, natural position. The ideal work position is to have the arms hanging relaxed from the shoulders. If a keyboard is used, arms should be bent at right angles at the elbow, with the hands held in a straight line with forearms and elbows close to the body. The head should be in line with the body and slightly forward.

Display Screens

When work is conducted at a computer, the top of the display screen should be at, or just slightly below, eye level. This allows the eyes to view the screen at a comfortable level, without having to tilt the head or move the back muscles.

Control glare at the source whenever possible; place VDTs so that they are parallel to direct sources of light such as windows and overhead lights and use window treatments if necessary. When glare sources cannot be removed, seek appropriate screen treatments such as glare filters. Keep the screen clean.

Your Chair

The chair is usually the most important piece of furniture that affects user comfort in the office. The chair should be adjusted for comfort, making sure the back is supported, and that the seat pan is at a height so that the thighs are horizontal, and feet are flat on the floor.

An ergonomically sound chair requires four degrees of freedom - seat pan tilt, backrest angle, seat height, and backrest height. Operators can then vary the chair adjustments according to the task. In general, chairs with the most easily adjustable dimensions permit the most flexibility to support people's preferred sitting postures.

Armrests on chairs are recommended for most office work except where they interfere with the task. Resting arms on armrests is a very effective way to reduce arm discomforts. Armrests should be sufficiently short and low to allow workers to get close enough to their work surfaces, especially for tasks that require fixed arm postures above the work surface.

Working Height

The work surface height should fit the task. The principle is to place the surface height where the work may be performed in such a manner as to keep arms low and close to the body in relation to the task.

If the working height is too high, the shoulders or the upper arms must be lifted to compensate, which may lead to painful symptoms and cramps at the level of the neck and shoulders.

If, on the other hand, the working height is too low, the back must be excessively bowed, which may cause backache. Generally, work should be done at about elbow height, whether sitting or standing.

Adjustable workstations should be provided so that individuals may change the stations to meet their needs. A VDT workstation without an adjustable keyboard height and without an adjustable height and distance of the screen is not suitable for continuous work.

Work/Rest Schedules

One solution for stress and fatigue is to design the computer operator's work so that tasks requiring concentrated work at the terminal are alternated with non-computer-based tasks throughout the workday. Also, a short break (5-10 minutes) should be taken at least once each hour when involved in continuous work at the computer.

Other Solutions

Additional measures that will aid in reducing discomfort while working with VDTs include:

- Change position, stand up, or stretch whenever you start to feel tired.

- Use a soft touch on the keyboard and keep your shoulders, hands, and fingers relaxed.

- Use a document holder, positioned at about the same plane and distance as the display screen.

- Rest your eyes by occasionally looking off into the distance.

Office Lighting

Different tasks require various levels of lighting. Areas in which intricate work is performed, for example, require greater illumination than warehouses. Lighting needs to vary from time to time and person to person as well. One approach is to use adjustable task lighting that can provide needed illumination without increasing general lighting.

Task lamps are highly effective to supplement the general office light levels for those who require or prefer additional light. Some task lamps permit several light levels. Since task lamps are controlled by the individual, they can accommodate personal preferences.

Indoor Air Quality

Indoor air quality (IAQ) is an increasingly critical issue in the work environment. The study of indoor air quality and pollutant levels within office environments is a complex problem. The complexity of studying and measuring the quality of office environments arises from several factors including:

- Office building floor plans are frequently changing to accommodate increasingly more employees and reorganization.

- Office buildings frequently undergo building renovations such as installation of new carpet, modular office partitions and free-standing offices, and painting.

- Many of the health symptoms appearing are vague and common both to the office and home environment.

- In general, little data on pollutant levels within office environments is available.

- Guidelines or standards for permissible personal exposure limits to pollutants within office buildings are limited.

Many times, odors are associated with chemical contaminants from inside or outside the office space, or from the building fabric. This is particularly noticeable following building renovation or installation of new carpeting.

Outgassing from such things as paints, adhesives, sealants, office furniture, carpeting, and vinyl wall coverings is the source of a variety of irritant compounds. In most cases, these chemical contaminants can be measured at levels above ambient (normal background) but far below any existing occupational evaluation criteria.

Waste Disposal

Office personnel should carefully handle and properly dispose of hazardous materials, such as broken glass. A waste receptacle containing broken glass or other hazardous material should be labeled to warn maintenance personnel of the potential hazard.

Office Chemical Safety

Each office employee must be made aware of all hazardous materials they may contact in their work area. The company's Hazard Communication Program includes:

- Written program.
- SDS for each hazardous substance used.

- Specific safe handling, use, and disposal guidance; and

- Employee training.

Emergency Action Plan

The company's Emergency Action Plan is designed to control events and minimize the effects of anticipated emergency situations. Through careful pre-planning, establishment of emergency action teams, training, and drills, employees can be safeguarded and potential for damage to company assets can be minimized.

The company Emergency Action Plan includes:

- Exits routes, meeting areas, and employee accounting.

- Emergency evacuation, incident command, and notification to emergency services.

- Personal injury and property damage.

- Protection of company information, both hard copy and electronic media.

- Bomb threats and facility security.

- First aid response.

- Use of fire extinguishers.

Emergency action team members (for example, supervisors, receptionist/telephone operators, and key assigned members) should be trained with quarterly reviews and drills. Semiannual drills with all employees should be conducted to ensure effectiveness. First aid kits or first aid supplies should be available with trained first aid providers available.

22 PERSONAL PROTECTIVE EQUIPMENT

Purpose

PPE requirements are determined by <u>documented PPE assessments</u> for each facility area or specific task.

Job Safety Analysis is an effective tool that can be used to determine necessary PPE for each job step. They are also effective in identifying hazards that can lead to safe jib procedures.

These requirements must be communicated to employees assigned to these areas and/or tasks. Required PPE may include the following:

- Approved leather safety toe boots.
- Long sleeve fire retardant clothing for welding and cutting.
- Approved hard hats where potential head injury is present.
- Eye and Face Protection

All eye and face protection required by this instruction must meet the requirements of ANSI 287.1

23 POWERED INDUSTRIAL TRUCK

Purpose

Material handling is a significant safety concern. During the movement of products and materials, there are numerous opportunities for injuries and property damage.

Powered industrial trucks, better known as forklifts, pallet jacks and stand-up riding reach trucks are essential tools in handling materials.

This policy has been created to minimize the risk of injury to operators, bystanders, and to avoid damaging property.

Qualified trainers will be used to provide all training activities.

Written records will be kept documenting all training.

Policy

Departments assigned powered industrial trucks must ensure that supervisors and operators comply with all aspects of this safety program.

All affected employees must successfully complete this training program and receive certification prior to the operation of any powered industrial truck.

Scope

This program applies to the operation of all powered industrial trucks, forklifts, tractors, platform lift trucks, motorized hand trucks, and other specialized industrial trucks powered by electric motors or internal combustion engines by employees and contractors, engaged in company projects.

Forklift Procedures

Pre-Use Inspection

- Prior to the operation of any powered industrial truck the Pre-Use Inspection Checklist must be completed. This applies at the beginning of every work period, and whenever a new equipment operator takes control of the powered industrial truck.
- Any safety defects (such as hydraulic fluid leaks; defective brakes, steering, lights, or horn; and/or missing fire extinguisher, lights, seat belt, or back-up alarm) must be reported for immediate repair. They must also be locked and tagged and taken out of service.

Operation

- Operators must always wear seat belts.
- Operators must sound the horn and use extreme caution when meeting pedestrians making turns, and cornering.
- Passengers are not allowed to ride on an industrial truck, unless the truck has an extra seat that allows the passenger to buckle-up while riding.
- Arms or legs may not be placed between the uprights of the mast or outside the running lines of the truck.
- Persons are not allowed to stand or pass under any elevated portion of a truck.
- Travel-ways must be maintained free from obstructions, aisles must be marked, and wide enough (six-foot minimum) for vehicle operation.
- Maintain sufficient headroom under overhead installations such as: lights, pipes, sprinkler systems, etc.
- An overhead guard must be used as protection against falling objects.
- Lift capacity must be marked on all powered industrial trucks. Operators must assure the load does not exceed rated weight limits.
- When a powered industrial truck is left unattended (more than 25ft. away or out of sight), load engaging means must be fully lowered, controls neutralized, power shut off, and brakes set. Wheels must be blocked if the truck is parked on an incline.

All modifications must be approved by the manufacturer, and new rated load capacities determined and posted on the truck. Written approval is required.

24 POWER & HAND TOOL SAFETY

General Requirements

- All hand and power tools, electrical cords, and pneumatic hoses shall be maintained in a safe condition.
- Faulty or damaged tools, cords, and hoses shall be tagged "Do Not Use" and removed from service immediately.
- Always use the proper tool for the job.
- Cheaters and other such devices to increase a tools capacity are not acceptable. TOOLS MADE FOR HANDLE EXTENSIONS ARE EXCLUDED.
- When power operated tools are designed to accommodate guards, they shall be equipped with such guards when in use.
- Cords and hoses shall be protected from damage and shall be routed through the job location, such that they are not tripping hazards.
- Employees using hand or power tools and exposed to the hazard of falling, flying, abrasive, or splashing objects or exposed to harmful dust, fumes, mist, vapors, or gases shall be provided with the proper personal protective equipment.

Tools

- Electric power operated tools shall either be of the approved double-insulated type or have cords which have the third wire ground whole and in place.
- Double insulated tools shall be clearly marked.
- Pneumatic power tools shall be secured to the hose or whip by some positive means to prevent the tool from becoming accidentally disconnected.
- Tools shall not be hoisted or lowered by their hoses.
- When fuel powered tools are used in enclosed places, the applicable requirements for concentration of toxic gases and use of PPE as outlined in 29 CFR 1910 shall apply.
- Power actuated tools shall be operated only by employees who have been trained in the operation of the tool in use.
- Hand tools shall be kept in good condition - sharp, clean, oiled, dressed, and not abused.
- Tools subject to impact tend to "mushroom" and shall be kept dressed to avoid flying spills.
- Wooden handles of tools shall be kept free of splinters and cracks and shall be kept tight in the tool.

Bench Grinders

Minimum Required PPE

- Safety Glasses
- Face Shield
- Hearing Protection
- Closed-Toe Shoes

Hazards & Controls

- High-speed abrasive wheels with a large amount of energy can be a contact hazard. Do not touch the wheel.

- Improper installation of the wheel or a worn-out wheel can lead to malfunction, becoming an explosion hazard and flying object hazard. Visual inspections, proper training of installation, and ring testing can prevent this from happening.

- Rotating parts leave potential for entanglement hazards with loose clothing, hair, or jewelry. Do not wear loose clothing or jewelry in the shop. Tie long hair up and back.

- Workpieces being kicked back by rotating wheel create flying object hazards. All objects should be secure on the work rest when being ground.

- Rotating parts can create a pinch point or crushing hazard. Do not touch the wheel. Keep all guards in place and at the proper distances.

- The grinding wheel is hot after use, creating a burn, heat, or fire hazard. Do not touch the wheel after it has been used.

- The grinding wheel generates sparks, creating a fire hazard. Keep all combustibles away from sparks.

- There is a possibility of dust exposure that may create a health hazard. Know the hazards that may be created by the work material.

- Dust and material build-up can easily be a housekeeping issue. Always keep work areas clean from dust and other foreign materials.

Limitations

- Bench Grinders should not be used for softer materials such as (but not limited to) non-ferrous metals (brass, aluminum, copper), plastics, or wood.

- Materials smaller than 3 inches cannot be ground unless a proper fixture is used to keep hands far enough from meeting the grinding wheel. The fixture must be able to hold the material being ground securely and without risk of becoming a projectile hazard.

Machine Guarding

All machine guards must be in place and/or properly Adjusted Before Use.

Other Precautions

- New grinding wheels should be installed and replaced by competent persons.

- The maximum RPM rating of the grinding wheel must be compatible with the RPM rating of the grinder motor.

- All grinding wheels should be visually inspected, and ring tested before first use and periodically afterward.

25 RESPIRATORY PROTECTION

Purpose

To define the selection, use, and care of respiratory protection.

Engineering Controls

- Removing or containing the hazard must be attempted before respirator use is considered.

- Respirator use is permitted only when engineering controls are not feasible, while engineering controls are being installed, or in emergencies.

Administrative Controls

- Evaluate the chemical products that workers must use.

- Ensure that workers use only that product which presents the least respiratory hazard and still does the job.

- Ensure that the Time Weighted Average for worker exposure does not exceed the Permissible Exposure Limit.

Selection

- Selection of respirators should be made according to OSHA 1910.134(d) – Selection of Respirators

- Only NIOSH/MSHA Approved Respirators May Be Used

- Employee Acceptance of a Particular Respirator Model Will Be Considered

  o Discomfort

  o Breathing resistance

  o Respirator weight

  o Interference with vision or work

Physical Fitness Determination

- Workers must be physically and physiologically capable of performing the assigned work and of wearing the selected respirator.

- Each employee who is required to wear a respirator must be evaluated by the company physician. The physician will rate the employee's ability to wear a respirator.

- Unless approval is granted by a company physician, no employee will wear a respirator except for emergency escape.

- If an employee refused to complete the questionnaire, Management will be contacted.

Instruction

- Each employee required to wear a respirator will be trained in the use and limitations of that respirator.

- Instruction will be given for every respirator.

- The instruction session will provide the employee with the opportunity:

  o To handle the respirator

  o To have it fitted properly.

  o To test its face to mask seal

  o To wear it in normal air for a sufficient time to become familiar with it

  o To wear it in a test atmosphere

- Instruction will be repeated, at least, annually.

Fit Testing

- A qualitative or quantitative respirator-fitting test will be used to determine the ability of each individual to obtain a satisfactory fit with a negative pressure respirator.

- All employees required to wear a negative pressure respirator will have each respirator they will use fit tested by one of the approved protocols.

- Under no circumstances will any employee be allowed to use any respirator if the results of the fit test indicate that the employee is unable to obtain a satisfactory fit with that respirator.

The Safety Department will be contacted if none of the available respirators will properly fit an employee who is required to wear a respirator.

No employee will be fit tested if beard stubble, sideburns, mustache, or other facial hair interferes with the face to facepiece seal or valves.

No employee will be fit tested for a full facepiece respirator if prescription eyeglasses are required for that person to safely perform the work or to see visual warnings given by a signal man.

The fit test portion of the Medical Questionnaire/Fit Test Form will be completed. The following must be recorded:

- The respirator models and sizes tested.
- Both successful and unsuccessful tests
- The protocols used.
- Date
- Instructor
- Employee's signature
- Fit tests will be repeated at least annually.
- More frequently if there is a change in facial configuration.

Respirator Use

- Manufacturer's recommendations for the proper use of respirators will be followed.

- The wearer will inspect the respirator immediately before donning.

- Defective respirators will be returned.

- Field fit checking will be performed each time a respirator is donned.

- Negative pressure respirators will be fit checked by negative pressure or positive pressure seal checks as recommended by the manufacturer.

- Pressure demand respirators will be fit checked by checking for air leaking around the facepiece.

- No employee will enter the hazardous area until the respirator is properly fitted.

- If an employee experiences any difficulty with a respirator, that person will leave the hazardous area immediately.

- Any difficulty fitting or using the respirator will be immediately reported to supervision.

- No employee will remove the respirator in the hazardous area until that area has been certified to be free of dangerous levels of contaminants.

IDLH atmospheres

- At least one standby person shall be present in the safe area.

- The standby will have the proper equipment available to assist the respirator wearers in case of emergency.

- Adequate and clear communication will be maintained between the standby and the respirator wearers.

- Employees wearing respirators in IDLH atmospheres will have a safety harness and lifeline or equivalent provision for removing them to the safe area in case of an emergency.

Respirator Cleaning, Storage and Repair

- Respirators will be cleaned and sanitized after each use.

- Strong cleaning and sanitizing agents and many solvents can damage respirator parts. These agents must be used with caution.

- High temperatures (>120-degree F) may damage respirator parts.

- Respirators may be washed in a detergent solution and sanitized with the following in a 2-minute immersion:

  o A hypochlorite (bleach) solution of 50 PPM

  o An aqueous iodine solution of 50 PPM

- o A quaternary ammonia solution 200 PPM
  (adjusted to the hardness of the water)

- Sanitizing solutions shall be completely rinsed from the respirator.

- Each respirator shall be inspected to determine if it is in good working order.

- Respirators shall be stored in a manner that will protect them from dust, sunlight, extreme heat or cold, excessive moisture, or damaging chemicals.

- Respirators shall be stored in a manner that does not distort the facepiece.

- Repairs to respirators shall only be made by qualified personnel.

26 SCAFFOLD SAFETY

General Requirements

All scaffolds shall be inspected before each use.

All scaffolds must be capable of withstanding four (4) times the maximum intended load except for suspended scaffolds which must be capable of withstanding six (6) times the maximum intended load.

Each platform is to be fully decked between the uprights and the guardrails as follows:

- Space between units is not more than one (1) inch except where it can be demonstrated that a wider space is necessary.

- Where more than one (1) inch is necessary, the platform shall be decked as tightly as possible with no more than a 9 ½ inch space between the platform and the uprights.

- Each end of the platform must extend at least six (6) inches over the support or be secured from movement by hooks or cleats.

- Each end of the platform must not extend more than twelve (12) inches over the support.

Access is required when platforms are more than twenty-four (24) inches above or below the point of access.

Access choices include portable ladders, stairways, stairway type ladders, ramps, walkways, integral prefabricated scaffold access or direct access.

- Cross braces shall not be used for access.

Fall protection is required on a scaffold more than ten (10) feet above a lower level

- Fall protection may consist of guardrail systems or personal fall arrest systems.

- When working on suspension scaffolds individuals must be protected by both guardrail systems and personal fall arrest systems.

Guardrails shall consist of a top-rail and a mid-rail and should be sufficient to hold 200 pounds. When the load is applied the top-rail shall not deflect below 38 inches.

Cross bracing may be used as a mid-rail if the crossing point of the two (2) braces is between 20 inches and 30 inches above the platform.

Cross bracing may be used as the top-rail if the crossing point at the two braces is between 38 inches and 48 inches above the platform.

Falling Object Protection

Each employee working on or around a scaffold shall be protected from falling objects by one of the following methods.

The area below is to be barricaded and employees not permitted to enter. Employees that must enter shall wear a hard hat.

A 4" toe-board is to be installed on the edge of the platform.

If the materials are piled higher than the toe-board, paneling or screening extending from the toe-board to the top of the guardrail shall be installed.

A canopy structure, debris net or catch platform is to be installed.

Training Requirements

No employee shall be permitted to do any of the following until he/she has received training from a qualified person.

- Scaffold Erection

- Scaffold Disassembly

- Scaffold Moving

- Scaffold Operating

- Scaffold Repairing

- Scaffold Maintenance

- Scaffold Inspection

All employees involved with the use of the scaffold shall be trained on the following:

- The nature of any electrical, fall and falling object hazard.

- The correct procedures for dealing with electrical hazards.

- The correct procedures for erecting, maintaining and disassembling fall protection systems and falling object protection systems.

- The proper use of the scaffold and the proper handling of materials on the scaffold.

- The maximum intended load and the load carrying capacities of the scaffolds used.

- The correct procedures for erecting, disassembling, moving, operating, repairing, inspecting and maintaining scaffolds.

- Any other pertinent requirements.

27 TRENCHING AND EXCAVATIONS

Application

This standard applies to all open excavations made in the earth's surface. Excavations are defined to include trenches.

NOTE: All trenches are excavations.

General Requirements

A competent person shall be assigned to all excavation work. This person shall be clearly identified to all employees assigned to the job.

Utility companies or owners must be contacted to establish the location of the utility underground installations before an excavation can begin.

All surface encumbrances must be removed or supported, as necessary, to protect employees.

A stairway, ladder, ramp or other safe means of egress shall be located in excavations that are four (4) feet or more in depth so as to require no more than 25 feet of lateral travel for employees.

No employee shall be permitted underneath loads handled by lifting or digging equipment. Employees shall be required to stand away from any vehicle being loaded or unloaded.

A warning system must be used when mobile equipment is used near an excavation and the operator does not have a clear and direct view of the edge.

Where oxygen deficiency or a hazardous atmosphere exists or could reasonably be expected to exist, the atmospheres in excavations four (4) feet or more in depth shall be tested.

Employees shall not work in excavations with accumulated water or in excavations in which water is accumulating.

Where the stability of adjacent structures is endangered by excavation operations, support systems such as shoring, bracing or underpinning shall be provided to ensure the stability of such structures and for employee protection.

Adequate protection shall be provided to protect employees from base rock or soil that could fall or roll from an excavation face. Protection may include scaling, barricades or other means that provide equivalent protection.

DAILY INSPECTIONS of excavations, the adjacent areas, and protective systems must be made by a competent person for evidence of possible cave ins, indications of failure of protective systems, hazardous atmospheres, or other hazardous conditions. The inspection must be prior to start of work and as needed throughout the shift. Inspection results must be recorded on "Competent Person" daily checklist.

Inspections also must be made after every rainstorm or other hazard-increasing occurrence. If the competent person finds evidence of a hazardous condition after an inspection, employees must be removed from the hazardous areas until necessary precautions are taken for protection.

Physical barriers shall be placed around or over excavations. Barriers shall be removed only when necessary to provide access to personnel or equipment. Flashing light barricades shall be provided at night.

Walkways or bridges shall be provided when employees or equipment are required or permitted to cross over excavations. Guardrails shall be provided where walkways are six (6) feet or more above lower levels.

Back-filling and removal of support systems shall be performed only after employees have cleared the area. All excavations shall be backfilled and graded promptly.

Walls and faces of excavations, four (4) or more feet deep, shall be shored, sloped or shielded as dictates by the type of soil encountered. Except when:

a.   Excavations are made entirely in stable rock.

b.   Excavations are less than five (5) feet in depth and examination of the ground by a competent person provides no indication of a potential cave in.

The protective systems shall have the capacity to resist without failure all loads that are intended or could reasonably be expected to be applied or transmitted to the system.

Soil Classification

Each soil and rock deposit at an excavation site must be classified by a competent person as stable rock, Type "A", Type "B" or Type "C" soil.

Stable rock means natural solid mineral matter that can be excavated with vertical sides and remain intact while exposed.

Type "A" soil is cohesive soil with an unconfined compression strength of 1.5 ton per square foot (TSF) (144Kpa) or greater. Cohesive soils include clay, sandy clay, clay loam and in some cases, silty clay loam and sandy clay loam. Cemented soils such as caliche and hard pan are also considered Type "A".

No soil should be classified as Type "A" if it is fissured; subject to vibration from traffic, pile driving, or similar effect; previously disturbed or part of a sloped, layered system where the slope is four horizontal to one vertical or greater.

Type "B" soil is cohesive soil with an unconfined compressive strength greater than 5 TSF but less than 1.5 TSF. Type "B" soils include granular cohesion less soils like angular gravely silt, silt loam, sandy loam and sometimes silty clay loam and sandy clay loam; previously disturbed soils except those which would otherwise be classed as Type "C" soil; fissured soils and soils subject to vibrations that would otherwise be classified as Type "A"; dry rock that is not stable; and material that is part of a sloped, layered system where the layers dip on a slope less steep than four (4) horizontal to one (1) vertical.

Type "C" soil is cohesive with an unconfirmed compressive strength of 5 TSF or less. Type "C" soils include granular soils such as gravel, sand, and loamy sand; submerged soil, soil from which water is freely seeping; submerged rock that is not stable; or material in a sloped, layered system when the layers dip into the excavation at a slope of four (4) horizontal to one (1) vertical or steeper.

Soil classification is to be based on at least one visual analysis and one manual analysis of the soil. Details of the acceptable visual and manual analysis may be found in Appendix "A" of the excavation standard.

28 VEHICLE SAFETY PROGRAM

Purpose

This program covers safe operation and maintenance of all company vehicles except those company vehicles regulated by the Interstate Commerce Commission (ICC) or US Department of Transportation DOT). Examples of vehicles covered include company-owned or leased passenger vehicles, pick-up trucks, light trucks, and vans that do not require drivers to possess a commercial driver's license (CDL) to operate.

Policy

- All company vehicles will be operated only by employees authorized by company management for specific company purposes.
- Vehicles will be maintained in a safe condition at all times. In the event of an unsafe mechanical condition, the vehicle will be immediately placed out-of-service and the appropriate manager notified.
- Only qualified company vehicle mechanics or approved service facilities are permitted to perform maintenance on company vehicles.
- All vehicles will be operated, licensed, and insured in accordance with applicable local, state, and federal laws.

- All employees authorized to operate any company owned or leased vehicle will be included in the company random drug testing program.
- All authorized employees must possess a valid state driver's license for the class of vehicle authorized.
- Authorized employees must have a driving record at least equal to that required for maintaining a CDL.

Responsibilities

Management

- Provide annual defensive driver training for all employees authorized to operate company vehicles.
- Train authorized employees on vehicle inspection and accident procedures.
- Maintain company vehicles in a safe condition.
- Maintain active insurance policies on all company vehicles.

Supervisors

- Allow only authorized employees to operate company vehicles.
- Arrange for defensive driving training prior to initial authorization.
- Maintain a list of authorized employees in their department.
- Arrange for required periodic maintenance checks on assigned vehicles.

- Immediately remove from service any vehicle with any safety defect.
- Not allow operation of any company vehicle by an authorized employee taking medication that warns of drowsiness.
- Establish a key control program for all assigned vehicles.

Authorized Employees

- Operate company vehicles in a safe, responsible manner and obey all traffic laws.
- Participate in driver training programs.
- Participate in the company drug testing program.
- Ensure all vehicle occupants use seatbelts before moving the vehicle.
- Follow safe fueling procedures.
- Conduct a pre-use inspection before any first daily use.
- Immediately report any safety defects or vehicle problems.
- Report use of all prescription medication.

Training

All employees authorized to operate company-owned or leased vehicles will participate in initial and annual driver safety training that will include:

- Defensive driving
- Vehicle inspection
- Accident procedures
- Hazardous weather driving
- Procedure for notification of unsafe vehicle
- Backing procedures (light truck and van operators)
- Cargo area storage (light truck and van operators)
- Loading and unloading (light truck and van operators)

Vehicle Inspection

Driver Inspections

Prior to each first daily use the driver shall inspect the vehicle for proper operation of the following safety features, as applicable:

- Horn
- Backup warning
- Head, tail, and signal lights
- Windshield wipers
- Tire inflation (visual check)
- Brakes
- Steering control
- Mirrors

- No operational warning lights
- Accident kit in glove compartment
- Fire extinguisher (light trucks and vans)
- Broken glass

Mechanical Inspections

Every company vehicle will be inspected by a qualified vehicle mechanics at least every three months. Inspection and maintenance points include:

- Road test
- Visual inspection of brake system (wheel removal required)
- Fluid system levels and visual inspection
- Brake pad wear
- Belts and hoses
- Battery condition
- Filter replacement
- Lubrication
- Oil change
- Emissions systems visual inspection
- Tire tread

All vehicle inspections and maintenance records will be maintained.

Driving Safely

Starting

- Conduct pre-use inspection.

- Always use seatbelts.
- Adjust seat and mirrors before starting vehicle.
- Allow a 15-second warm up time.
- Check for warning lights.

Driving

- Do not drive if drowsy.
- Think ahead - anticipate hazards.
- Don't trust the other driver to drive properly.
- Don't speed or tailgate.
- Drive slower in hazardous conditions or hazardous areas.
- Pass only in safe areas and when excessive speed is not required.
- No loose articles on floor
- Do not read, write, apply make-up, drink, eat or use a phone while driving.
- Stay at least four seconds behind the vehicle ahead.
- Do not stop for hitchhikers or to provide roadside assistance.

Backing

- Back slowly and be ready to stop.
- Do not back up if anyone is in path of vehicle travel.
- Check clearances.
- Don't assume people see you.
- Get out and check if you cannot see from the driver's seat.

Stopping

- Park only in proper areas, not roadsides
- Use warning flashers and raise hood if vehicle becomes disabled.

Accidents

- Notify your company and law enforcement as soon as possible.
- Cooperate with any law enforcement officers.
- Move the vehicle only at the direction of a law enforcement officer.
- Fill out all sections of the accident report in the glove box.
- Do not sign any forms unless required by a law enforcement officer.
  - o At the scene get the following information
  - o Investigating officer name and law enforcement agency
  - o Make, model, and license plate number of other vehicles.
  - o Names, addresses, and phone numbers of all witnesses
  - o Photos of accident using camera in glove box:
- Name, address, and license number of other driver(s)

29 WELDING, CUTTING, AND BRAZING

General Requirements

- A hot work permit may be required for welding, cutting, and brazing operations when required. Hot work permits shall be issued by the safety coordinator where practical.

- Suitable fire extinguishing equipment shall be immediately available in all welding, cutting and brazing work areas.

- A fire watch shall be provided as required by location procedures and shall be maintained for at least 30 minutes after completion of the job.

- Objects to be welded, cut, or heated shall be moved to a designated safe location, or if they cannot be readily moved, all moveable fire hazards in the vicinity shall be taken to a safe place. If fire hazards cannot be removed, positive means shall be taken to confine the heat, sparks, and slag and to protect the immovable fire hazards from them.

- Spark containment shall be utilized during all welding, burning, and grinding operations. Spark containment may include laying fire blankets, placing barricades, totally enclosing the spark producing operation, or by the use of fire watch. Employees working around or below the welding, burning, or grinding operation shall be protected from falling or flying sparks.

- Welding, cutting and heating may normally be done without mechanical ventilation or respiratory equipment, but where an unsafe accumulation of contaminants exists, suitable mechanical ventilation or respiratory protective equipment shall be provided.
- Whenever welding, cutting or heating is performed in a confined space, exhaust ventilation shall be provided. When sufficient ventilation cannot be provided, employees shall be protected by airline respirators.

Gas Welding and Cutting

- All hoses, torches, and/or bottles carrying acetylene, oxygen, fuel gas or any substance which may ignite or be harmful to employees shall be inspected at the beginning of each working shift. Defective hoses and torches shall be tagged "DO NOT USE" and immediately removed from service.
- Torches shall be lighted from friction lighters and not by matches or from hot work.
- Directional gas flow fittings (back-flow valves) shall be provided on hoses to prevent reverse gas flow or back flow (LOCATED AT TANKS AND TORCH BODY)
- Torches shall be turned off and removed from confined spaces when not in use.

Transporting, Moving, and Storing Compressed Gas Cylinders

- Valve protection caps must be in place and secured, whenever cylinders are not in use. They may not be used to lift cylinders from one vertical position to another.
- Cylinders must be secured on a cradle, sling board, or pallet when hoisted. They may not be hoisted or transported by means of magnets or choker slings.
- Cylinders should be moved by tilting and rolling them on their bottom edges. They should not be intentionally dropped, struck, or permitted to strike each other violently.
- When cylinders are being transported by a powered vehicle, the cylinder must be secured in a vertical position.
- A suitable cylinder truck chain or other steadying device must be used to keep cylinders from being knocked over while they are being used. Cylinders must be secured in an upright position at all times.
- The cylinder valve must be closed when work is finished, when cylinders are empty or when cylinders are moved at any time.
- Compressed gas cylinders must be stored in an upright position at all times.
- When oxygen cylinders are stored, they must be separated from fuel-gas cylinders or combustible materials (especially oil or grease) either a minimum

distance of 20 feet (6.1m) or by a noncombustible barrier at least 5 feet (1.5 m) high, having a fire-resistance rating of at least one-half hour.

- When stored inside a building, cylinders must be placed in a well-protected, well ventilated, dry location, at least 20 feet (6.1m) from highly combustible materials, such as oil or excelsior. Cylinders must be stored in assigned places, away from elevators, stairs, or gangways. These assigned storage places must be located where cylinders will not be knocked over or damaged by passing or falling objects or be subject to tampering by unauthorized persons. Cylinders must not be kept in unventilated enclosures, such as lockers and cupboards.

Use of Fuel Gas and Proper Instruction in Use

- Before connecting a regulator valve to a cylinder, the valve must be opened slightly and closed immediately ("cracked") to clear the valve of dust or dirt that might otherwise enter the regulator. The person cracking the valve must stand to one side of the outlet, not in front of it. When cracking the valve of a fuel gas cylinder, care must be taken to ensure that the gas does not reach welding work, sparks, flame, or other possible sources of ignition.
- Opening and closing of cylinder valves:
  o The cylinder valve must be opened slowly to prevent damage to the regulator.

- o Valves on fuel gas cylinders must be opened no more than 1.5 turns, so that they may be closed quickly when necessary.
- o If a special wrench is required to close a valve, it must be left in position on the stem of the valve while the cylinder is in use so that the gas flow can be shut off quickly in the event of an emergency.
- o For manifold or coupled cylinder, at least one such special wrench must always be available for immediate use.
- o While a fuel gas cylinder is being used, nothing that might damage a safety device or interfere with the quick closing of the valve may be placed on top of the cylinder.
- If fuel gas from a cylinder is being used in a torch or other device equipped with a shut off valve, a regulator attached to the cylinder valve or manifold must be used to reduce the pressure.
- Before a regulator may be removed from a cylinder valve, the cylinder valve must be closed, and the gas released from the regulator.

Fuel Gas and Oxygen Manifolds

- Gas must be clearly identified on the manifold.
- Fuel gas may not be placed in enclosed spaces.
- A hose may not be of a type that can be interchanged between fuel gas, oxygen manifolds, and supply header connections.

Arc Welding and Cutting

- Arc welding and cutting operations shall be shielded by non-combustible or flameproof screens which will protect employees and other persons working in the vicinity from the direct rays of the arc.
- Arc welding and cutting cables shall be of the completely insulated, flexible type, capable of handling the maximum current requirements of the work in progress. Cables in need of repair shall not be used.
- When the welder or cutter has occasion to leave work or to stop work for any appreciable length of time, or when the welding or cutting machine is to be moved, the power supply switch to the equipment shall be shut down.
- All ground return cables and all arc welding and cutting machine grounds shall be in accordance with 29 CFR 1910 and ground connections shall be made directly to the material being welded.

30 EMPLOYEE SAFETY AGREEMENT

I have read (or the rules have been read to me) and I understand the safety rules of this organization.

I agree to take responsibility for my own safety and the safety of those around me by complying with all local, state, and federal regulations, as well as the safety rules of this organization.

I understand that these safety rules do not constitute any form of binding promise or contract for this organization to continue to employ me for any specific period or under any specific circumstances.

I also understand that this organization may change its rules if it so chooses.

NAME (printed): _____

SIGNATURE: _____

DATE: _____